MIT Press Series on Economic Learning and Social Evolution
General Editor
Ken Binmore, Director of the Economic Learning and Social
Evolution Centre, University College London.

1. *Evolutionary Games and Equilibrium Selection*, Larry Samuelson, 1997

Evolutionary Games and Equilibrium Selection

Larry Samuelson

The MIT Press
Cambridge, Massachusetts
London, England

First MIT Press paperback edition, 1998

© 1997 Massachusetts Institute of Technology

This book was set in Palatino by Windfall Software using ZzTeX.

Printed and bound in the United States of America.

Library of Congress Cataloging-in-Publication Data

Samuelson, Larry, 1953–
  Evolutionary games and equilibrium selection / Larry Samuelson.
    p. cm. — (MIT Press series on economic learning and social
  evolution)
  Includes bibliographical references and index.
  ISBN 0-262-19382-5 (hc: alk. paper), 0-262-69219-8 (pb)
  1. Game theory. 2. Noncooperative games (Mathematics)
3. Equilibrium (Economics) I. Title. II. Series.
HB144.S26 1997
519.3′02433—dc20   96-41989
                                                                   CIP

To my parents

Contents

Series Foreword

The MIT Press series on Economic Learning and Social Evolution reflects the widespread renewal of interest in the dynamics of human interaction. This issue has provided a broad community of economists, psychologists, philosophers, biologists, anthropologists, and others with a sense of common purpose so strong that traditional interdisciplinary boundaries have begun to melt away.

Some of the books in the series will be works of theory. Others will be philosophical or conceptual in scope. Some will have an experimental or empirical focus. Some will be collections of papers with a common theme and a linking commentary. Others will have an expository character. Yet others will be monographs in which new ideas meet the light of day for the first time. But all will have two unifying features. The first will be a rejection of the outmoded notion that what happens away from equilibrium can safely be ignored. The second will be a recognition that it is no longer enough to speak in vague terms of bounded rationality and spontaneous order. As in all movements, the time comes to put the beef on the table—and the time for us is now.

Authors who share this ethos and would like to be part of the series are cordially invited to submit outlines of their proposed books for consideration. Within our frame of reference, we hope that a thousand flowers will bloom.

Ken Binmore
Director
Economic Learning and Social Evolution Centre
University College London
Gower Street
London WC1E 6BT, England

Preface

This book began as simply a collection of my recent articles in evolutionary game theory. As the book has evolved, I have done some rewriting of these articles in an attempt to make clear the connections between them and the common themes that run through them. At the same time, I have endeavored to retain the ability of each chapter to stand alone.

There are already several good texts or surveys, listed at the beginning of chapter 1, devoted to evolutionary game theory. In light of these, I do not attempt a comprehensive treatment of the subject in this book. A brief review is given of the background material that is particularly relevant, some guidance is given as to where a more detailed treatment can be found in the literature, and then the book concentrates on my work.

The articles forming the basis for this work have been written with Ken Binmore, John Gale, Georg Nöldeke, Richard Vaughan, and Jianbo Zhang. I am fortunate to have had the opportunity to work with these coauthors and I have learned much from them. The especially large role that Ken Binmore and Georg Nöldeke have played will be apparent in the pages that follow. Any remaining errors are my responsibility.

I am thankful to Ken Binmore, George Mailath, Phil McCalman, Georg Nöldeke, and two anonymous reviewers for helpful comments and discussions on this manuscript, and I am thankful to Bonnie Rieder for drawing the figures. Part of this work was done while visiting the CentER for Economic Research at Tilburg University, the Departments of Economics at the Universities of Alicante, Bonn and Melbourne, and the Institute for Advanced Studies at the Hebrew University of Jerusalem. I am grateful to each of them for their support and hospitality. Financial support from the National Science Foundation (SES 9122176 and SBR 9320678) and the Deutsche Forschungsgemeinschaft (SFB 303) is gratefully acknowledged.

1 Introduction

This book examines the interplay between evolutionary game theory and the equilibrium selection problem in noncooperative games. It is motivated by the belief that evolutionary techniques have much to tell us about equilibrium selection. The analysis focuses on this theme, leaving to others the task of presenting more comprehensive coverage of either evolutionary game theory or equilibrium selection.[1]

This chapter provides an introduction to equilibrium selection and evolutionary models beginning with the most basic of questions, including what game theory is and what one might hope to accomplish with it. A discussion of these seemingly elementary issues may appear unnecessary, but I am convinced that paying careful attention to such questions is important to making progress in game theory.

1.1 Game Theory

Game theory is the study of *interactive* decision making, in the sense that those involved in the decisions are affected by their own choices *and* by the decisions of others. This study is guided by two principles. First, people's choices are motivated by well-defined, stable preferences over the outcomes of their decisions. Second, people act strategically. When making their decisions, they take into account the relationship between their choices and the decisions of others.

The assumption that people have stable preferences, and make choices guided by these preferences, has long played a central role in

---

1. For book-length introductions to evolutionary game theory, see Bomze and Pötscher [42], Fudenberg and Levine [95], Hofbauer and Sigmund [119], Maynard Smith [149], Vega-Redondo [242], and Weibull [248]. A more concise introduction appears in van Damme [240]. The survey articles of Crawford [67] and Mailath [145] are also helpful. For a discussion of equilibrium selection, see van Damme [241].

economics. This feature is often identified as the defining characteristic of economic inquiry. Contrary to the second principle, however, much of economic theory is constructed on the assumption that people do *not* act strategically. Instead, the links between an agent's decisions and the decisions of others are assumed to be sufficiently weak as to be safely ignored, allowing the agent to act as if he is the isolated inhabitant of an unresponsive world.[2] The consistent exploitation of this assumption leads to one of economic theory's most elegant achievements, general equilibrium theory, as well as to one of its most commonly used, partial equilibrium supply-and-demand analysis.

That game theory and conventional or "nonstrategic" economic theory are both built upon preference-based individual decision theories ensures that they have much in common. The nonstrategic approach to economics has been quite successful. It has passed the market test of being used, both within and outside of academia. I believe that, whenever possible, game theory should be guided by the same principles of inquiry that have worked for conventional economics.

Three characteristics of a conventional economic model are noteworthy. First, the heart of such a model is an equilibrium. The equilibrium provides a convenient tool for organizing the information contained in the model, even if we do not believe that the world is always in equilibrium. Second, the usefulness of the model generally lies in its ability to yield comparative static results showing how this equilibrium changes as parameters of the model change. Finally, behind this equilibrium analysis lie implicit assumptions about out-of-equilibrium behavior, designed to address the questions of how an equilibrium is reached or why the equilibrium is interesting.

In equilibrium, the agents that inhabit a traditional economic model are rational, flawlessly making choices that maximize their profits or utility. However, if one examines the out-of-equilibrium story that lurks behind the scenes, one often finds that these agents are anything but rational. Instead, their decisions may be driven by rules of thumb that have very little to do with optimization. However, market forces select in favor of those whose decisions happen to be optimal while eliminating those making suboptimal choices. The standard defense of profit maximization, for example, is that firms employ a wide variety

---

2. It is perhaps for this reason that Robinson Crusoe often appears on graduate qualifying exams in microeconomic theory.

of pricing rules, but only those whose rules happen to provide good approximations of profit maximization survive. Similarly, consumers may behave rationally in a Walrasian equilibrium, but are typically quite myopic in the adjustment process that is thought to produce that equilibrium. The behavior that persists in equilibrium then looks as if it is rational, even though the motivations behind it may be quite different. An equilibrium does not appear because agents are rational, but rather agents appear rational because an equilibrium has been reached.

As successful as the nonstrategic approach has been, some economic problems involve strategic considerations in such a central way that the latter cannot be ignored. For example, consider the discussion of markets that commonly appears in intermediate microeconomics textbooks. As long as one side of the market has sufficiently many agents that each individual can approximately ignore strategic considerations, then traditional economic theory encounters no difficulties. It is straightforward to discuss perfect competition, monopoly, and monopsony. But once these three have been considered, an obvious fourth case remains, namely bilateral monopoly, as well as intermediate cases involving markets in which the number of buyers or sellers exceeds one, but is not large. There is no escaping strategic considerations in these cases, and no escaping a need for game theory.

How can we use game theory to address strategic issues? There are two possibilities. For some, game theory is a normative exercise investigating how decisions *should* be made. The players that inhabit this game theory are hypothetical "perfectly rational agents." The task for game theory is to formulate a notion of rationality, derive the implications of this notion for the outcomes of games, and then ensure that the resulting theory is consistent, in the sense that we would be willing to call the behavior that produces such outcomes rational. The intricacy of this problem grows out of the self-reference that is built into the notion of rationality, with all of its attendant logical paradoxes. The richness of the problem grows out of our having no a priori notion of what it means to be rational, substituting instead a mixture of intuition, analogy, and ideology. The resulting models have a solid claim to being philosophy and art as well as science.[3] This version of game theory

---

3. Game theory has experienced considerable success as a normative pursuit. For example, one can find discussions of moral philosophy (Binmore [20], Gauthier [98], and Harsanyi [114]) and social choice (Moulin [159]) carried out in game-theoretic terms.

departs from conventional economic inquiry in that rational behavior is the point of departure, rather than being the outcome of the adjustments that produce an equilibrium.

For others, game theory is a positive exercise in investigating how decisions *are* made. The players that inhabit this game-theoretic world are people rather than hypothetical beings. The goal is a theory providing insight into how people are observed to behave and how they are likely to behave in new situations. In taking people as its object of inquiry and taking insight into behavior as its goal, this use of game theory follows squarely in the footsteps of traditional economic theory.

We shall be concerned with the positive uses of game theory. It is therefore important to be clear on what is being sought from game theory. I do not expect the formal structure of the theory, by itself, to have positive implications. Instead, I expect to derive implications from the combination of game theory and additional assumptions about players' utilities, beliefs, rules for making decisions, and so on. We can then never "test" pure game theory. Instead, game theory by itself consists of a collection of tautologies that neither require nor admit empirical testing. It is accordingly more appropriate to view game theory as a language than a theory. At best, we can test joint hypotheses that involve assumptions beyond game theory. The true test of game theory is not a statistical investigation of whether it is "right" but rather a more practical test of whether it is useful, just as the success of nonstrategic economic theories is revealed in their having been found to be useful. Those pursuing game theory as a positive exercise must demonstrate that it is a valuable tool for investigating behavior.

The example of traditional economic theory, and especially expected utility theory, is instructive in this respect. Utility maximization by itself has no positive implications for behavior. One can make any behavior the outcome of a utility maximization problem by cleverly constructing the model, primarily by choosing the appropriate preferences. In spite of this, utility maximization retains its place as the underlying model for virtually all of economics. This is partly because utility maximization provides a convenient language in which to conduct economic analysis. To a much greater extent, however, utility maximization retains its central place in economic analysis because the addition of some simple further structure, such as the assumption that people prefer more to less, yields a model with sufficient power to be useful.[4]

Game theory has experienced some success as a positive tool by providing a framework for thinking precisely about strategic interactions. Even seemingly simple strategic situations contain subtle traps for the unwary, and even the clearest of thinkers can find it difficult to avoid these traps in the absence of game-theoretic tools to guide their thinking. At an abstract level, familiar choice problems such as Newcomb's paradox, the surprise test paradox, and the paradox of the twins are less paradoxical when one has used the tools of game theory to model the strategic situation (see Binmore [20]). In economic applications, game theory has provided the tools for understanding a wide variety of behavior. For example, it was once common to construct models of limit pricing in which incumbent firms manipulated their current prices to deter new firms from entering the market, even though, as observed by Friedman [84], there was no link between current prices and entrants' postentry profitability. An understanding of these issues appeared only with the use of game-theoretic techniques (e.g., Milgrom and Roberts [154]). Porter [184] uses game theory to describe the price-setting behavior of American railroads in the 1880s; Hendricks, Porter, and Wilson [117] find game-theoretic models useful in explaining the behavior of bidders in auctions for offshore oil and gas drilling rights; and Roberts and Samuelson [191] use game theory to describe the advertising behavior of American cigarette manufacturers.

1.2 Equilibrium

Along with some successes, game theory has encountered difficulties. As with nonstrategic analyses, a game-theoretic investigation typically

---

4. The observation might be made here that utility maximization at least implies the generalized axiom of revealed preference (GARP), and we can check for violations of GARP, either by observing actual choices or conducting experiments. However, there is a sufficiently large leap from the theoretical world where GARP is derived to even the most tightly controlled experimental or empirical exercise as to allow sufficient room for an apologist to rationalize a failure of GARP. For example, the choices that we can observe are necessarily made sequentially, which in turn causes different choices to be made in slightly different situations and with slightly different endowments, allowing enough latitude for apparent violations of GARP to be explained. The attraction of expected-utility theory arises because the additional assumptions needed to give it predictive power, such as that small changes in endowments do not matter too much, often appear plausible.

proceeds by formulating a model and then finding an equilibrium, in this case a Nash equilibrium.[5] However, many games have more than one equilibrium, and for an equilibrium analysis to be successful, we must have some way of proceeding in such cases. Making progress on the problem of equilibrium selection is essential to the success of game theory.

This is not to say that we must choose a *unique* equilibrium for every game. Instead, there may be some games in which our best guess is that the outcome will be one of several equilibria. If the game is a coordination game in which people are deciding whether to drive on the right or the left side of the road, for example, then it seems pointless to argue about which side it will be without seeking further information. Simply given the game, our best statement is that one of these outcomes will appear. This is consistent with the evidence, as we observe driving on the right in some countries and driving on the left in others. We could construct a theory explaining which countries drive on the right and which on the left, but the most likely places to look for explanations include factors such as historical precedents and accidents that lie outside of the description of the coordination game commonly used to model this problem. If we have only the information contained in the specification of the game, our best "solution" is to offer both equilibria.

This may suggest that we resolve the equilibrium-selection problem simply by offering the set of all Nash equilibria as solutions. This is an approach often taken by nonstrategic economic theories, where multiple equilibria also arise, but recourse can be found in showing that all of the equilibria share certain properties. For example, the fundamental welfare theorems establish properties shared by all Walrasian equilibria, reducing the need to worry about selecting particular equilibria. In game-theoretic analyses, however, this approach usually does not work. Instead, some equilibria appear to be obviously less interesting than others. In the driving game, for example, we readily dismiss the mixed-strategy equilibrium in which each driver drives on each side of the road with equal probability. As the following paragraphs show, there are other games in which some pure-strategy equilibria are commonly dismissed.

---

5. A Nash [164] equilibrium is a collection of strategies, one for each player, that are *mutual best replies* in the sense that each agent's strategy is optimal given the strategies of the other agents.

	A	R
N	0, 4	0, 4
E	2, 2	−4, −4

Figure 1.1
Chain-Store Game

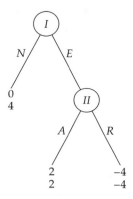

Figure 1.2
Chain-Store Game, extensive form

Much of the recent work in game theory has been concerned with the question of what to do when there are multiple equilibria, with particular attention devoted to characterizing properties that might help distinguish equilibria worthy of our attention from less interesting equilibria. To illustrate the issues, consider the game shown in figure 1.1.[6] It is most convenient to think of this game in its extensive form, shown in figure 1.2. This is a one-period version of Selten's Chain-Store Game [209]. Player *I* is a potential entrant into a market. If player *I* stays out of the market (*N*, or "Not enter"), then player *II* is a monopolist and earns a payoff of 4. If player *I* enters (*E*, or "Enter"), then player *II* can acquiesce (*A*), causing the two firms to share the market for profits of 2 each, or player *II* can retaliate (*R*), causing a price war that drives both players' payoffs to −4.

The Chain-Store Game has two pure-strategy Nash equilibria, given by (*E*, *A*) and (*N*, *R*). Traditional game theory typically endorses (*E*, *A*) and rejects (*N*, *R*). The difficulty with the latter is that it calls for

---

6. The convention is adopted throughout that player *I* is the row player and player *II* the column player.

player II to retaliate if the entrant enters, even though retaliating gives player II a lower payoff than acquiescing. Notice that retaliating is a best reply for player II in the Nash equilibrium (N, R), because the mere "threat" of retaliation induces player I to stay out and ensures that player II does not actually have to carry through on this threat. But this threat does not appear to be "credible," in the sense that player II would prefer to not retaliate if player I actually chose E.

Not only do some equilibria appear to be less plausible than others, such as (N, R) in the Chain-Store Game, but the implications of the analysis can depend crucially on which equilibrium we choose. I have suggested limit pricing as a case in which our understanding was significantly enhanced by the use of game theory. But Milgrom and Roberts's game-theoretic analysis of equilibrium limit pricing [153] yields multiple equilibria. Entry is affected by limit pricing in some of these equilibria, but not others. Some firms distort their preentry prices in some equilibria, but not others. This is precisely the type of multiple equilibrium problem that we must address if game theory is to be a useful, positive tool. Is the absence of entry in a market an indication that entry is not profitable, or that incumbents are taking anticompetitive actions? What are the welfare effects of the actions firms take in response to potential entry, and what should be our policy toward these actions? Insight into these questions hinges upon which of the many possible equilibria we expect to appear.

The initial response to the problem of multiple equilibria, the equilibrium refinements literature, refined the definition of Nash equilibrium by adding additional criteria. These criteria are designed to exclude equilibria such as (N, R) in the Chain-Store Game, in the process offering precise specifications of what it means for a "threat" not to be "credible."

The difficulty with (N, R) in the Chain-Store Game appears to be obvious. In the normal form of the game, R is a weakly dominated strategy.[7] In the extensive form, the equilibrium is supported by a threat of responding to entry by playing R—a threat that player II would prefer not to carry out should entry occur. Equivalently, when player I

---

7. It is becoming increasingly common to use the phrase "strategic form" rather than "normal form," perhaps in response to a fear that "normal form" carries with it the presumption that this is the preferred way to examine a game. "Normal form" is a shortening of von Neumann and Morgenstern's "normalized form" [245, p. 85], which does not signal a priority for examining such games; von Neumann and Morgenstern identify the "normalized form" as being better suited to some purposes and the "extensive form" to others. I use "normal form" throughout.

chooses N, player II has a collection of alternative best replies available, including the pure strategies A and R, all of which maximize player II's payoffs. The equilibrium (N, R) is supported by a particular choice from this collection that appears to be the wrong choice.

The initial building blocks of equilibrium refinements were stipulations that dominated strategies must not be played or that choices at out-of-equilibrium information sets must be equilibrium choices in the counterfactual circumstance that these information sets are reached. For example, one of the first and most familiar refinements, Selten's notion of a subgame-perfect equilibrium [207, 208], is built around the requirement that equilibrium play occur even at out-of-equilibrium subgames. The effect of such requirements is to place restrictions on which strategy from a collection of alternative best replies can be used as part of a Nash equilibrium, in the process eliminating equilibria such as (N, R) in the Chain-Store Game.

The restrictions to be placed on out-of-equilibrium behavior may appear to be obvious in the Chain-Store Game, but more formidable challenges for an equilibrium refinement are easily constructed, involving more complicated out-of-equilibrium possibilities as well as more intricate challenges to our intuition as to what is a "good" equilibrium. The result has been a race between those who construct examples to highlight the flaws of existing refinements and those who construct refinements. This race has produced an ever-growing collection of contending refinements, many of which are quite sensitive to various fine details in the construction of the model, but the race has produced very little basis for interpreting these refinements or choosing between them.[8]

The possibility that different refinements will choose different equilibria is only one of the difficulties facing the refinements program. In other games the conventional refinements agree, only to produce a counterintuitive outcome, where the refinements appear to miss some important ideas about how games are played. Consider the Coordination Game shown in figure 1.3. There are again two pure-strategy Nash equilibria, given by (T, L) and (B, R). These are both strict Nash

---

8. The equilibrium refinements literature has been successful in forcing game theorists to make precise their arguments for choosing one equilibrium over another. Intuition and judgment may still be the best guide to equilibrium selection, but refinements provide the means for identifying and isolating the features of an equilibrium that make it intuitively attractive. One of the first refinements of sequential equilibrium was named the "Intuitive Criterion" by Cho and Kreps [64] in recognition of its being designed to capture what they found intuitively appealing about certain equilibria.

	L	R
T	2, 2	0, 0
B	0, 0	1, 1

Figure 1.3
Coordination Game

	L	R
T	0, 0	1, 1
B	1, 1	0, 0

Figure 1.4
Coordination Game

equilibria and hence involve no alternative best replies. As a result, both equilibria survive virtually all of the equilibrium concepts in the refinements literature. Yet we would be surprised to see anything other than (T, L) from this game.

In the course of concentrating on evaluating Nash equilibria in which some players play weak best replies, such as (N, R) in the Chain-Store Game, the refinements literature has avoided the relative merits of strict Nash equilibria.[9] The choice between such equilibria is often described as a separate question of equilibrium *selection* rather than refinement. Kohlberg and Mertens [135, note 2] go so far as to maintain that the choice between strict Nash equilibria lies outside the purview of noncooperative game theory. However, I prefer not to separate the question of choosing an equilibrium into two separate pieces, using a refinement theory to deal with Nash equilibria involving alternative best replies and a selection theory to deal with choices between strict Nash equilibria. Instead, I join Harsanyi and Selten [116] in thinking that behavioral considerations relevant to one of these issues must be relevant to the other as well. The title of this book accordingly refers only to equilibrium *selection* (as does the title of Harsanyi and Selten's).

In yet other games, there are often no Nash equilibria that appear to be a likely description of behavior. Consider the Coordination Game shown in figure 1.4. This game has two pure-strategy Nash equilibria, given by (T, R) and (B, L), and a mixed-strategy equilibrium in which each player attaches probability $\frac{1}{2}$ to each pure strategy. Which equilib-

---

9. The most notable exception is Harsanyi and Selten [116].

rium should we expect? In the absence of a clear answer to this question, do we have any reason to believe that players confronted with this game will manage to coordinate their actions on a pure-strategy Nash equilibrium, or on any equilibrium? In some cases, as in the choice of which side of the road to drive on, historical precedent or something in the description or environment of the game may focus players' attention on a particular Nash equilibrium. In other cases, there may no be obvious clues as to how to play the game. The Nash equilibrium concept then seems to be too strong.[10]

In response to such concerns, Bernheim [14] and Pearce [182] chose the opposite direction of the refinements literature, introducing the concept of "rationalizable" strategies.[11] In figure 1.4 all strategies are rationalizable. More generally, any strategies that are part of a Nash equilibrium are rationalizable, but there may be other rationalizable strategies. Rationalizability is designed to capture precisely the implications of the common assumptions that the players know the game, are Bayesian rational, and all of this is common knowledge, and it is accordingly significant that the implications of rationalizability fall short of Nash equilibrium.[12]

In some cases, it may even be too demanding to insist on rationalizable strategies. For example, the only rationalizable outcome in the finitely repeated Prisoners' Dilemma is to defect in every round, but we do not always expect relentless defection. The apparent lesson to be learned here is that the common knowledge of rationality is a very demanding assumption. There may be cases where players do not know even that their opponents are rational, much less that this is common knowledge, and such cases may well take us outside the set of rationalizable strategies.[13]

---

10. One might appeal to the mixed-strategy equilibrium in this case, but this raises the question of why players should use or believe in precisely the required mixtures.

11. A strategy for player *I* is rationalizable if it is a best response to some belief about player *II*'s strategy that puts positive probability only on player-*II* strategies that are best responses to some belief about player *I*'s strategy that puts positive probability only on player-*I* strategies that are best responses and so on.

12. A flourishing literature has grown around the question of rationalizable strategies, including work on alternative notions of rationalizability, especially in extensive-form games (Bernheim [14], Battigalli [10, 11], Börgers [45, 46], and Pearce [182]) and work using the concept of rationalizability in applications (Cho [63], Watson [247]).

13. Kreps and Wilson [137] and Milgrom and Roberts [154] exploit the lack of common knowledge of rationality to explain entry deterrence in Selten's Chain-Store Game [209]. Kreps et al. [136] analogously explain cooperation in the finitely repeated Prisoners' Dilemma. Fudenberg and Maskin [96] provide a similar argument for general finitely repeated games.

The difficulties that arise in choosing between equilibria, or in deciding whether *any* equilibrium is applicable, reflect a deeper lack of consensus among game theorists on how to interpret a Nash equilibrium. One interpretation is that players are to be viewed as having no idea who their opponents are or how the game has been played in the past, either by themselves or others, and are locked into separate cubicles that isolate them from any observations of the outside world. Each must choose a strategy while in this cubicle and without any knowledge of others' choices, though perhaps armed with the common knowledge that all players are rational. Here, it is hard to see how we could impose any requirement stronger than rationalizability. In particular, it is hard to see how we can expect an equilibrium in figure 1.4 in such circumstances.

Alternatively, the Nash equilibrium concept is often motivated by assuming that the players have *unlimited* opportunities to communicate, but then must independently choose their strategies (perhaps in the cubicles of the previous paragraph). The presumed outcome of this communication is an agreement on how the game should be played. Here, it is generally assumed that a Nash equilibrium agreement must result: the players will understand that if they agree on strategies that are not a Nash equilibrium, then at least one of them will have an incentive not to honor the agreement. It is generally further assumed that once an agreement on a Nash equilibrium has been struck, the fact it is a best reply for each player to perform her part of this agreement suffices to ensure the agreement will be executed.[14] If this motivation for Nash equilibrium is persuasive, however, then it would seem to have much stronger implications than simply Nash equilibrium or various of its refinements. For example, it is hard to imagine the equilibrium (B, R) emerging in figure 1.3 in such circumstances.

1.3 Evolutionary Games

The one constant that runs throughout conventional game theory, including especially the equilibrium refinements and equilibrium selec-

---

14. What happens in the absence of an agreement is left unspecified. But this is important, because the communication phase itself is a game that may well have multiple equilibria as well as many nonequilibrium outcomes. An interesting device for surmounting this difficulty is to replace the conversation and resulting agreement with a recommendation made by an exogenous authority. One might think of this authority as being "Game Theory." Greenberg's theory of social situations [106] is motivated in this way.

tion literature, is the belief that players are rational, and this rationality is common knowledge. The common knowledge of rationality is often informally regarded as a necessary condition for there to be any hope that equilibrium play will appear. However, this same rationality immediately creates tension for an equilibrium theory. Players use their equilibrium strategies because of what would happen if they did not. But rational players are certain to play their equilibrium strategies. One is then forced to confront counterfactual hypotheses of the type: what happens if a player who is sure to play his equilibrium strategy does not do so?[15]

Philosophers have long wrestled with the meaning of counterfactuals. Lewis [141] discusses the meaning of counterfactuals in terms of "possible worlds." No player will deviate from her equilibrium strategy in this world, meaning the world in which the equilibrium is indeed an equilibrium, but there are other possible worlds in which she would deviate. She does not deviate because doing so would have adverse consequences in these other worlds. For example, we can be confident that in this world, we will not leap off the edge of a tall building because there are possible worlds in which we can imagine taking such a leap and in which the consequences are undesirable.

Considerable care must be taken in contemplating the other worlds required to evaluate counterfactuals. We can readily imagine possible worlds in which we leap off a building and fall to destruction. However, there are yet other possible worlds in which a leap off a tall building does not have undesirable consequences, perhaps because the laws of gravity work somewhat differently in these worlds. We are persuaded not to leap, in spite of being able to imagine these latter worlds, because the worlds with undesirable consequences somehow seem "closer" to our own world than do those with benign consequences. Our task in evaluating counterfactuals is to find the *closest* possible world to our own in which the deviation giving rise to the counterfactual occurs. An equilibrium is a collection of strategies that are optimal when the consequences of deviations are evaluated in the closest possible world.

The difficulty here is that there is no objective way of determining which possible worlds are closest. Logic alone makes no distinction between the world in which we leap to our deaths and the world in which we happily float in midair. Instead, an equilibrium selection

---

15. The discussion in section 1.3 is based on Binmore and Samuelson [23].

theory assigns meaning to the concept of closest as part of the process of determining how out-of-equilibrium moves are to be evaluated.

Where should we look for insight into which alternative possible worlds are close? This again depends upon our view of game theory. If game theory is a normative exercise, this amounts to asking what a rational player should think when confronted with an impossibly irrational play. There appears to be very little to guide our thinking in this respect, though equilibrium concepts such as justifiable equilibrium (McLennan [152]) and explicable equilibrium (Reny [185]) are built around versions of requirements that agents should believe their opponents are as rational as possible.

If game theory is a positive exercise, then it is important to realize that the models with which we work are simplified representations of more complicated real-world problems. The obvious place to look when considering close alternative worlds is then the features of the actual problem that have been assumed away in constructing the model. The world is rife with imperfections that we ignore when constructing models, and an equilibrium concept implicitly chooses among these imperfections to explain out-of-equilibrium behavior.

Assumptions about closest worlds are typically built into equilibrium concepts in the form of trembles. The original game is replaced by a new game that matches the original with high probability, but that with some small probability is actually a different game. Any action that should not have occurred in an equilibrium of the original game is then explained as indicating that the players are really playing the altered game, which is carefully chosen so that any such action is consistent with the play of the game. Selten's trembling hand perfection [208], for example, adds chance moves to the original specification of the game so that players intending to choose an equilibrium action play that action most of the time, but occasionally play something completely different. Faced with an action that would not occur in equilibrium, an agent assumes that Nature intervened to replace the intended equilibrium action with the actual action.

Economists are often reluctant to abandon the perfectly rational optimizing agents that frequent economic models. The trembles upon which an equilibrium refinement is built are often introduced with an appeal to the possibility that players might make mistakes, or that one can only understand "complete rationality as a limiting case of incomplete rationality" (Selten [208, p. 35]). However, the resulting trembles typically affect the rules of the game (as in Selten's trembling hand perfection), the preferences of the players, or the beliefs of the players. The common practice is to ask a player to stubbornly persist in the belief that her opponent is rational, even after observing a tremble that would provide good evidence to the contrary if trembles really represented departures from rationality.

The point of departure for an evolutionary model is the belief that people are not always rational. Rather than springing into life as the result of a perfectly rational reasoning process in which each player, armed with the common knowledge of perfect rationality, solves the game, strategies emerge from a trail-and-error learning process in which players find that some strategies perform better than others. The agents may do very little reasoning in the course of this learning process. Instead, they simply take actions, sometimes with great contemplation and sometimes with no thought at all. Their behavior is driven by rules of thumb, social norms, conventions, analogies with similar situations, or by other, possibly more complex, systems for converting stimuli into actions. In games that are played rarely and that are relatively unimportant, there will rarely be any discipline brought to bear on such behavior; we will have little to say about them. In games that are sufficiently important to attract players' attention, however, and that are played repeatedly, players will adjust their behavior, rejecting choices that appear to give low payoffs in favor of choices that give high payoffs. The result is a process of experimenting and groping about for something that seems to work, a process in which strategies that bring high payoffs tend to crowd out strategies that do not.

In adopting this approach, evolutionary game theory follows the example set by nonstrategic economic models. Nonstrategic equilibrium theories insist on rationality when describing their equilibria, but resort to less-than-rational behavior when telling the stories that motivate equilibrium. Evolutionary game theory similarly assumes that the behavior driving the process by which agents adjust their strategies may not be perfectly rational, even though it may lead to "rational" equilibrium behavior. In particular, the adjustment process driving evolutionary game theory sounds much like the process by which competitive markets are typically described as reaching equilibrium, with high-profit behavior being rewarded at the expense of low-profit behavior. The evolutionary approach thus marks not a new departure for economists but rather a return to their roots.

Like traditional game theory, evolutionary models again produce a tremble-based theory of equilibrium, but with a new notion of tremble. If a player makes a choice that would not be made in equilibrium, the most likely tremble to which an evolutionary game theorist appeals is that the player's reasoning does not match the reasoning specified by the equilibrium analysis. If we are involved in a game of chess and believe that play is proceeding according to some equilibrium, what should we believe when our opponent makes an unexpected move? Traditional game theory calls for us to believe that the opponent meant to make the equilibrium move, but Nature intervened in its execution, perhaps by causing the player's hand to slip. Alternatively, we might believe that the player has different beliefs about the rules of the game or preferences over the outcome. Evolutionary game theory would suggest that we consider the possibility that the player has reasoned differently about the game and has some other path of play in mind.[16]

The example of chess provides some interesting insights into the interplay between evolutionary game theory and equilibrium. Chess is a finite, zero-sum game of perfect information. As a result, either White has a pure strategy that forces a victory (i.e., that allows While to win no matter how Black plays), or Black has a pure strategy that forces a victory, or each player has a pure strategy that forces a draw.[17] Which of these is the case is not yet known, and chess is accordingly still an interesting game, but a perfectly rational player would know which was the case. A perfectly rational player then need never draw any inferences about her opponent. Her strategy prescribes an action for every contingency, and she need only pursue this strategy, without giving any thought to her opponent, to ensure the outcome she can force. If the opponent is rational, there is no point in doing anything else. Against the type of opponent found in evolutionary game theory, however, it may pay quite well to do something else. Suppose the solution to chess is that each player can force a draw. Let the players begin with strategies that force this outcome, but let Black make an unanticipated move. This move may reveal an unexpected way of thinking about the game that will lead to more unorthodox moves in the future. It may also be that White will draw if White simply continues with her strategy, but will win if, after recognizing the implications of Black's unexpected move, she plays in a way that exploits Black's future play,

---

16. This alternative path of play may be characterized as a mistake on the part of the opponent, meaning that the opponent would do better to stick to the equilibrium we had in mind, or perhaps more usefully, as a mistake on our part, meaning that the supposed "mistake" on the part of the opponent is simply an indication that the path we have in mind does not provide a good model of our opponent's behavior.
17. This result dates back to Zermelo [254] and is often identified as the beginning of game theory.

even though such attempted exploitation would be disastrous against a rational Black. The failure to act accordingly puts one in the situation of the "unlucky expert" who loses at poker or bridge or chess, only to explain that he would have won if only his opponents had played better.[18]

1.4 Evolution

How do we translate these ideas about equilibrium and out-of-equilibrium play into tools that can be used to analyze games? There is no biology in this book, and yet the word "evolution" is used throughout. Evolutionary game theory takes both its name and its basic techniques from work in biology. In a biological application, the players in the game are genes; the strategies are the characteristics of behavior with which these genes endow their host organism. The payoff to a gene is the number of offspring carrying that gene. The link between strategies and payoffs arises because some characteristics will make their host organisms better able to reproduce than others.

In some cases, the payoff to a gene will depend only on its own characteristic or behavior and will not depend upon the genes carried by other organisms. This is the case of *frequency-independent selection* and is the counterpart of a nonstrategic approach in economics. In other cases, the success of one gene will depend upon the other genes present in the population of organisms. This is likely to occur whenever the host organisms must compete for resources such as food, shelter, or mates. This is the case of *frequency-dependent selection*, and it is here that evolutionary game theory is especially useful.

Evolutionary game theory proceeds in one of two ways. The original approach, pioneered by Maynard Smith and Price [150] and Maynard Smith [149], is to work with the concept of an *evolutionarily stable strategy* (ESS). An ESS is a Nash equilibrium satisfying an additional stability property. This stability property is interpreted as ensuring that if an ESS is established in a population, and if a small proportion of the population adopts some mutant behavior, then the process of selection arising out of differing rates of reproduction will eliminate the latter. Once an ESS becomes established in a population, it should therefore be able

---

18. The Connecticut Yankee in King Arthur's court (Twain [236]) is consistently led to ruin by assuming that his bumbling opposition is rational. He observes that the worst enemy of the best swordsman in France is by no means the second best swordsman in France but rather the worst swordsman in France because the latter's irrational play cannot be anticipated by one trained in the art of rational swordsmanship.

to withstand the pressures of mutation and selection. Biologists have found the ESS concept useful. A good example is provided by Hammerstein and Riechert [112], who show that the ESS concept provides a good description of the behavior of spiders when contesting ownership of territories.

The alternative is to construct a dynamic model of the process by which the proportions of various strategies in a population change. This will typically be a stochastic process, but if the random events affecting individual payoffs are independent, and the population sufficiently large, a good approximation is obtained by examining the expected value of this process. In a biological context, such approximations typically lead to the often-used replicator dynamics.[19] The best-known form of the replicator dynamics was introduced by Taylor and Jonker [232] and by Zeeman [253], but similar ideas have been used in earlier applications. Hofbauer and Sigmund [119] discuss one of the earliest analyses to use similar ideas, in which the mathematician Volterra [244] constructed a dynamic model to explain population fluctuations of predator and prey fish species in the Adriatic Sea. Similar equations had previously been studied by Lotka [144], and are now known as "the Lotka-Volterra equations."

The agents in biological applications of evolutionary game theory never choose strategies and never change their strategies. Instead, they are "hard-wired" to play a particular strategy, which they do until they die. Variations in the mix of strategies within a population are caused by differing rates of reproduction. In economic applications, we have a different process in mind. The players in the game are people, who choose their strategies and may change strategies.[20] An evolutionary approach in economics begins with a model of this strategy adjustment process. In some cases, this model lies implicitly behind the scenes and the analysis proceeds by applying static solution concepts to the game, such as ESS, that are designed to characterize the outcomes of

---

19. Dawkins [71, p. 15] introduces the term *replicator* to refer to a collection of molecules capable of producing a copy of itself, or replicating, using raw materials either from itself or from other sources. Hofbauer and Sigmund [119, pp. 147–148] note the central role played by the replicator dynamics in a variety of biological applications, including models of sociobiology, macromolecular evolution, mathematical ecology, and population genetics.

20. Dawkins [71] suggests a view of the evolution of human behavior in which the place of genes is taken by "memes," which are rules of thumb for making decisions. Dawkins interprets the study of evolution to include the process by which people make selections from the pool of memes.

	C	D
C	3, 3	0, 5
D	5, 0	1, 1

Figure 1.5
Prisoners' Dilemma

the strategy adjustment model. In other cases, an explicitly dynamic approach is pursued.

The use of evolutionary techniques in economics is often associated with Robert Axelrod [3] and his Prisoners' Dilemma tournaments. A typical Prisoners' Dilemma is shown in figure 1.5, where we interpret strategy C as cooperation and D as defection. The Prisoners' Dilemma has gained notoriety as a game in which we have a particularly compelling solution concept, namely strictly dominant strategies, yielding an outcome, (D, D), that is to some disconcertingly inefficient (because the nonequilibrium outcome strategies (C, C) give both players higher payoffs).

Allowing agents to play an infinitely repeated version of the Prisoners' Dilemma opens the possibility for cooperation as an equilibrium outcome.[21] However, the repeated game also opens the door to a multitude of other equilibria, including one in which players relentlessly defect. Which should we expect, and do we have any reason to expect cooperation to appear?

Axelrod [3] addressed this question by staging a tournament. Invitations were issued for game theorists and others to submit computer programs capable of playing the repeated Prisoners' Dilemma. Each program submitted played every other program in a round-robin tournament, including a game against itself and a game against an additional strategy that played randomly in each period. In the first of Axelrod's tournaments, fourteen submitted strategies played a version of the repeated Prisoners' Dilemma that lasted exactly two hundred rounds. The winner of the tournament, meaning the strategy with the highest payoff in the round-robin tournament, was the strategy TIT-FOR-TAT.[22]

---

21. Elements of cooperative outcomes in repeated games also appear in the biological literature under the guise of reciprocal altruism.
22. TIT-FOR-TAT cooperates on its first move and thereafter plays whichever strategy its opponent played in the previous period.

A second and better-known tournament attracted sixty-two submissions (in addition to including the random strategy). In this case, contestants were not told the length of the game, but were told that in each period there would be a probability of .00346 that the game would end after that period and a probability of .99654 that the game would continue. Axelrod made this stopping rule operational by taking five draws from the distribution over game lengths implied by these probabilities, producing lengths of 63, 77, 151, 156, and 308. Each match then consisted of five games, one of each of these lengths. The winner of this tournament was again TIT-FOR-TAT.

Axelrod followed this tournament with an evolutionary analysis. His interpretation was that if future tournaments were held, then less successful strategies would be submitted less often, and more successful strategies resubmitted more often. Axelrod accordingly constructed a dynamic model to simulate this strategy selection process. Each strategy began each period with a weight. Using these weights, Axelrod calculated the average payoff of each strategy against all other strategies. He then adjusted the weights, increasing the weight on those strategies with relatively high average payoffs and decreasing the weights on strategies with low average payoffs. These weights might be interpreted as the proportions of a population consisting of the various strategies, with a process of selection increasing the proportion of high-payoff strategies and decreasing the proportion of low-payoff strategies. When the simulation halted, TIT-FOR-TAT had the highest population proportion, about 15%.

The success of TIT-FOR-TAT was initially attributed to TIT-FOR-TAT's being an evolutionarily stable strategy for the infinitely repeated Prisoners' Dilemma, though it is now clear that the infinitely repeated Prisoners' Dilemma has no ESS. Linster [142, 143] and Nachbar [162] have analyzed and extended Axelrod's work; a growing list of studies simulate evolutionary models of the Prisoners' Dilemma.[23] Linster and Nachbar's work shows that the identity of the most successful strategy in the repeated Prisoners' Dilemma can be quite dependent upon the pool of available strategies, and identifies cases in which strategies other than TIT-FOR-TAT are most successful. Linster [142], for example, examining a model containing the twenty-six strategies that can

---

23. See, for example, Bruch [52], Huberman and Glance [122], Nowak [171], Nowak, Bonhoeffer, and May [172], Nowak and May [173, 174], Nowak and Sigmund [175, 176, 177, 178, 179], and Nowak, Sigmund, and El-Sedy [180].

be implemented by one- or two-stage finite automata, finds that if the evolutionary process is subjected to constant perturbations, then GRIM, which prescribes cooperation until the first instance of defection, after which it prescribes relentless defection, is the most successful strategy. Together with the simulations, Linster and Nachbar's work indicates that the repeated Prisoners' Dilemma is a complicated environment in which TIT-FOR-TAT sometimes but not always fares well.

1.5   Evolution and Equilibrium

The use of evolutionary game theory in economics has now spread well beyond the Prisoners' Dilemma. In the process, attention has turned to the implications of evolutionary processes for equilibrium selection. Do evolutionary models yield outcomes in which agents appear to act rationally, in the sense that they avoid dominated strategies or choose rationalizable strategies? Are the outcomes of evolutionary models Nash equilibria? Are evolutionary models helpful in choosing refinements of Nash equilibria, or in selecting between strict Nash equilibria?

Our approach to these questions will be characterized by four themes: (1) the use of dynamic models, (2) the consideration of the time span over which these models are relevant, (3) a focus on the stability properties of the outcomes, and (4) an interest in the equilibrium implications of these outcomes.

Dynamics

I work primarily with dynamic models. Whenever possible, I construct these dynamic models from explicit specifications of how individuals choose their strategies. It is tempting to claim that the result is less arbitrary models, but we have no nonarbitrary way of choosing some models as being less arbitrary than others. However, an advantage of an explicit foundation is that it forces us to be specific about the forces behind our model and result, just as equilibrium refinements brought precision to assessments of alternative equilibria. This appears to be our best hope for interpreting the analysis. Although it will do little to resolve questions about what is the right model, it may at least focus the discussion on the right questions.

The models employed in this book are extraordinarily simple. One cannot hope that such models literally capture the way people make decisions. Instead, like all models, they are at best approximations of how agents behave. Better approximations would be valuable, and it is important that we make progress on the question of how people learn to use certain strategies.[24] At the same time, it is important to investigate the implications of various learning rules, and especially the extent to which these implications are driven by broad properties of the learning rules rather than idiosyncratic details. Among other insights, this will help focus our attention on the interesting issues when examining how people learn.

There is very little that one would recognize as "rational" in the way the agents modeled in this book behave. One approach to adaptive behavior is to construct models in which agents are fully rational Bayesian expected-utility maximizers, but are uncertain about some aspect of their environment. Their behavior then adapts as new information prompts the updating of their expectations. We have much to learn from such models, but I suspect that real-world choices are often guided by quite different considerations. The starkly clear theoretical models with which we work may look much different to the people actually facing the situations approximated by these models. They may not even recognize that they are involved in a game and may not know what strategies are available or what their payoffs are, much less be able to form a consistent model of this situation and act as Bayesian expected-utility maximizers. Instead, I suspect that their actions will often be driven by imitation, either of others or of their own behavior in situations they perceive to be similar. Their behavior may then be quite mechanical, bearing little resemblance to that of a Bayesian utility maximizer facing uncertainty.

I describe the forces guiding agents' strategy choices as "selection" or "learning." I use these terms interchangeably, though the former has a more biological flavor, while the latter is commonly used in economic contexts. Given the mechanical behavior described in the previous paragraph, is it reasonable to characterize the agents modeled in this book as learning? A committed Bayesian would answer with a clear no. In order to be called learning, a Bayesian would demand a specification of what the agents are learning, and would then want to see this specification captured in a model of Bayesian updating. The

---

24. Psychologists have been concerned with such issues for a much longer time than have economists, and we undoubtedly have much to learn from the psychological learning literature.

behavior described in this book would perhaps be described as adaptation, but not learning.

If you were to replace "learning" with "adaptive behavior" throughout this book, I would not object. The key point is that people adapt their behavior to their circumstances. Is a model of Bayesian updating the best way to capture this behavior? Perhaps sometimes, but in many cases, Bayesian models divert attention from the important aspects of behavior. A Bayesian analysis requires the agents to begin with a consistent model of their environment, which can then be mechanically adjusted as new information is received. Attention is directed away from the process by which this consistent model is achieved and focused on the subsequent adjustments. However, I think the crucial aspects of the learning process often occur as agents attempt to construct their models of the strategic situation they face. These attempts may often fail to produce consistent models, with agents unknowingly relying upon a contradictory specification, until some event exposes the inconsistency and sends the agents scurrying back to the model construction stage. The models people use may often be so simple as to make the Bayesian analysis of the model trivial, but to make the model construction and revision process extremely important.

I do not have a model of how people construct the models that guide their behavior. Instead, I rely upon crude and often arbitrary specifications of how choices are made. However, if we are interested in people, rather than ideally rational agents, then I suspect that working with such specifications will be useful.

Time

The typical procedure in evolutionary game theory is to examine the limiting behavior of a dynamic model, either explicitly or implicitly, through an equilibrium concept such as an ESS. Is this an appropriate way to proceed?[25]

A good start on answering this question is provided by asking when an evolutionary model can be expected to be useful. Many of the difficulties with interpreting Nash equilibrium and its refinements arise out of a failure to identify the circumstances in which it is relevant, or out of attempts to apply a single concept in widely different circumstances. I have mentioned two common motivations for Nash equilibrium, one

---

25. This section expands upon a discussion in Binmore and Samuelson [26].

allowing virtually no coordination and one explicitly allowing for coordination. It is then no surprise that some object to the Nash equilibrium concept as magically assuming too much coordination, while others regard it as naively assuming too little.

It makes no sense to apply an evolutionary analysis to games that are played only occasionally, or to situations in which agents rarely find themselves. Behavior in such situations is likely to be driven by a host of factors we would traditionally identify as economically insignificant, including seemingly unimportant differences in the way people perceive the problem they must solve; evolutionary considerations will be rendered irrelevant by the lack of opportunity to adjust behavior in light of experience. I suspect that psychology has more to tell us about such situations than does economics.

It also makes no sense to apply an evolutionary analysis in cases where the game is complicated and the rewards from learning how to play the game optimally are small. In the course of our daily affairs, we make so many decisions that we cannot subject each one to a careful analysis. Many decisions are made with the benefit of very little consideration at all in order to concentrate on seemingly more important decisions. Little attention has been devoted to the question of which decisions command our attention and which do not, but we cannot expect people to devote resources to problems that are either too difficult to solve or not worth solving.

In the circumstances just described, the arguments against an evolutionary analysis are also arguments against an *equilibrium* analysis. It is hard to see how agents can be expected to coordinate on equilibrium actions in situations that are only rarely encountered, or in cases where players do not devote critical attention to the game. Where an evolutionary analysis does not apply, we can then hope, at best, to restrict attention to rationalizable strategies.

Given a case where an evolutionary model is useful, meaning a strategic situation that is frequently encountered and sufficiently simple and important to command attention, we must next ask how this model is to be used. In particular, we can examine the stationary distributions or steady states of the model, or we can examine the path along which the model travels on its way to such a stationary distribution.

Economists devote the bulk of their attention to long-run behavior. This is reflected in the tremendous energy devoted to studying the equilibria of models, and the very little to understanding the process that produces these equilibria. Even dynamic models are typically studied by examining the long-run behavior captured by a steady state analysis.

Evolutionary game theorists are similarly interested in the long run, although here it is less clear what is meant by the "long run." The extent to which a long-run analysis is interesting, and the relationship between the long-run outcome of a model and the behavior the model is designed to study, depends crucially on choices made when constructing the model. For example, Hammerstein and Riechert [112] are interested in the long-run behavior of spiders in territorial contests. But even without further modeling, we know the stationary distribution of the actual process governing the spiders' behavior, which is that the spiders are extinct. In particular, we know that with positive probability there will come a period of time when all spider offspring will be born male, and that the period will be long enough for this outcome to ensure extinction. In spite of this prediction, we continue to study spiders.[26] The long run in which we are interested is some period of time sufficiently short that such an event occurs only with extraordinarily small probability, but sufficiently long for other forces to have produced a stable pattern of behavior among the spiders. This is typically studied by constructing a model that excludes such unlikely events as an accidental extinction caused by a prolonged period of all-male births, and then studying the long-run behavior of this new, stripped-down model.

Studying the long-run behavior, or equivalently the stationary distributions, of the resulting model corresponds to studying part of the *path* along which the underlying system travels on its way to its stationary distributions. This path is relevant because the events that shape a stationary distribution of the underlying system occur with such low probability, and the expected time until such an event occurs is so long, that the intervening behavior is worthy of study.

We are thus interested in the paths along which behavior evolves. Which aspects of these paths are to be studied? We are unlikely to have much to say about the behavior characterizing the initial segments of such paths. In a biological system, this behavior will be determined by the initial genetic structure of the population, which in turn will have been determined by the past effects of selection on the population. In an economic context, initial behavior when faced with a new problem

---

26. A similar calculation predicts that the fate of the human race is extinction, but this has not halted social science.

is likely to be determined by rules of thumb or social norms that people are used to applying in similar but more familiar situations. Which situations are considered similar, and hence which rules are used, is likely to depend upon the way the new problem is described and presented, and may have little to do with the economics of the problem.

I use terms such as *rule of thumb, convention,* and *social norm* to describe the factors determining the initial behavior in a decision problem or game. Questions such as what constitutes a convention or norm, how one is established, how it shapes behavior, and how it might be altered, have attracted considerable attention. In contrast, I do not even define these terms, and frequently use them interchangeably. The important point is that behavior in a newly encountered strategic situation is often not the result of a careful strategic analysis. Many factors may influence this behavior, including especially the way the player would behave in similar, more familiar situations. These are the factors to which I refer when speaking of rules of thumb, norms, and conventions. I often use more than one term because the factors in question are likely to be quite varied. I refrain from a more thorough analysis of these factors because my interest lies primarily in what happens as they are displaced by strategic considerations.

If we allow sufficient time to pass, then the state of the system will respond to the forces of selection or learning. Given our currently primitive understanding of learning, it is not clear that we can expect evolutionary models to usefully describe the initial phases of this learning process, during which the rules of thumb that produced the initial behavior in the game contend with the economic considerations that players learn in the course of playing the game.

If we are lucky, the learning or selection process will approach a seemingly stationary state and remain in the vicinity of that state for a long time. It is in studying this aspect of a path that simple evolutionary models are most likely to be useful. At this point, the original rules of thumb and social norms guiding behavior will have given way to economic considerations. We may then be able to approximate the learning process with simple models sufficiently closely that the models provide useful information about the such states.

If we are unlucky, the learning dynamics may lead to cyclic behavior or more complicated behavior, such as chaos, that has nothing to do with stationary points. Although we know from work in dynamic systems that cyclic or complicated behavior is a very real possibility, appearing in seemingly simple models and excluded only at the cost of strong assumptions, I shall restrict my attention to sufficiently simple games that such behavior does not occur. This approach clearly excludes many games, but it seems wise to understand simple cases first before proceeding to more complicated ones. Similarly, a stationary distribution will exist in all of my models.

Even though we shall be examining models in which behavior approaches a stationary configuration, it is important to remember this is not the study of the stationary distributions of the underlying evolutionary process. These will typically depend upon the occurrence of rare events that have been excluded from the model, such as the possibility of a sex ratio realization that is sufficiently peculiar as to ensure extinction. For example, suppose we are interested in a coordination game in which drivers must decide whether to drive on the right or the left side of the road and must also decide upon a unit in which to measure the distances they drive. After some initial experience, we would expect some standard of behavior to be established, though there are several possibilities for such a standard, and we may not be able to limit attention to just one of these. People in the United States have coordinated on driving on the right side of the road and measuring their distances in miles; in England people drive on the left and measure in miles, while in France they drive on the right and measure in kilometers. Whichever standard of behavior is established, we would expect the system to linger in the vicinity of such behavior for an extended period of time. This may accordingly be a stationary distribution of the model we have constructed. However, this behavior need not characterize a stationary distribution of the underlying process. Countries have been known to change both the side of the road on which they drive and the units in which they measure.

When examining the sample paths of an evolutionary process, we may be working with either a stationary distribution of the model we have constructed or the sample paths of the model. If we have reason to believe that the expected waiting time until reaching a stationary distribution of the *model* is relatively short, then we are more likely to be interested in the stationary distribution. In other cases, however, reaching a stationary distribution of the model will require an extraordinarily long waiting time. This is likely to be the case if the model captures the events whose improbability leads to an interest in the *paths* rather than the stationary distribution of the underlying evolutionary process. In these cases, we will be especially interested in the sample paths of the model, including the states near which these paths linger for long periods of time.

Returning to the coordination game in which people must choose the side of the road on which they will drive, one possible model would consist simply of best-response dynamics. The sample path of such a model will lead to coordination on one of the two possible outcomes, and then will exhibit no further change. Faced with this model, we would be interested in stationary distributions. However, evolutionary game theorists have recently made great use of models in which an adjustment process, such as the best-response dynamics, is subjected to perturbations that are very rare but can cause large changes in behavior.[27] The sample path of such a model will again initially produce coordination on one of the two possible sides of the road, such as the right side, and will remain in the vicinity of such an outcome for a very long time. However, this state will not provide a good description of the stationary distribution of the model. At some point a perturbation will shift the model to the vicinity of the state in which everyone drives on the left, where the system will then linger for an extended time. Over the course of time, an infinite number of such shifts between sides of the road will occur, between which the system spends long periods of time when very little movement occurs. A stationary distribution is a distribution over states induced by such shifts. However, the time required to establish the stationary distribution of the model is likely to be extraordinarily long. In this case, we will often be interested in the sample path behavior of the model.

It will be helpful to have terms to describe the various phases of an evolutionary process. In Binmore and Samuelson [26], we referred to the initial behavior of the system as the "short run." We referred to the period of time during which learning is adjusting the strategies of the agents but before behavior has approached a state near which it will spend an extended period of time as the "medium run." Roth and Erev [195] use "medium run" similarly.

Kandori, Mailath, and Rob [128] and Young [250] use "long run" to refer to a stationary distribution of their processes in which best response dynamics are subjected to rare perturbations. However, these stationary distributions are driven primarily by large bursts of mutations that happen very rarely and for which we must wait a very long time. We have just seen that when faced with such a process, we may

27. For example, Kandori, Mailath, and Rob [128] and Young [250].

be most interested in studying the limiting outcome of a model from which these rare bursts of mutations are excluded. As shown in chapter 3, removing the perturbations from the underlying process can lead to the study of a system of differential equations. But studying the limiting outcome of such a model is also often described as a "long-run analysis" (as in Weibull [248]).

The phrase "long run" thus has two masters to serve. In Binmore and Samuelson [26], we responded by using "ultralong run" to refer to the stationary distributions of an evolutionary process, including the influence of possibly very rare events that may give rise to long waiting times before this stationary distribution is relevant. "Long run" was used to refer to a study of the limiting outcomes of a model from which the rare perturbations of the underlying process have been removed. The long run is then a period of time long enough for the process to have converged to a stable pattern of behavior. This behavior may not characterize a stationary distribution of the underlying process, but it will characterize the behavior of the process for a long time. The short, medium, and long runs are all sample path phenomena of the underlying process, while the ultralong run is characterized by a stationary distribution of that process, though the long run may be characterized by a stationary distribution of a *model* constructed to approximate the underlying process by excluding certain very unlikely events.

Stability

The primary technique for examining sample paths of the underlying model will be to approximate these paths by systems of differential equations and then to examine the stationary states of these equations. Does this provide a relevant guide to long-run behavior? This depends upon the stability properties of these stationary states. For example, mathematicians generally regard a stationary state that is a sink as having reassuring stability properties, while remaining skeptical of a saddle or a source.

The difficulty is that when working with games, we frequently encounter stationary states that are neither sinks, saddles, nor sources. In the language of the theory of differential equations, these stationary states are not hyperbolic.[28] In the eyes of a mathematician, to fail to

---

28. The Jacobian of the system of differential equations has eigenvalues with zero real parts at these states.

be hyperbolic is one of the most serious sins for a stationary state. Two difficulties arise with a nonhyperbolic stationary state. First, examining the eigenvalues of the associated Jacobian does not reveal the stability properties of the stationary state. It might or might not be a sink, and more information is required to determine which is the case. This is a signal that the first-order approximation captured by examining the Jacobian does not provide enough information to answer stability questions. More importantly, whatever the stability properties of this stationary state, they are not "structurally stable." Instead, an arbitrarily small perturbation in the specification of the system of differential equations can completely alter the properties of the stationary state. Some perturbations may turn the state into a sink, others into a source, and still others may destroy the stationary state altogether.

Mathematicians often respond by identifying nonhyperbolic stationary states as being uninteresting. To bolster this point of view, theorems have been developed identifying conditions under which "almost all" dynamic systems, in some suitable sense, have only hyperbolic stationary states. It is then no surprise that mathematicians commonly answer questions about nonhyperbolic stationary states with the question, "Why worry about such exceptional cases?"

Economists also often make use of genericity conditions. I agree that we should work with generic models. A result driven by a nongenericity places too much faith in our ability to capture exactly the strategic situation in a model. However, it is important that we work with an appropriate concept of genericity. For example, we often encounter economic questions that are best modeled as extensive-form games, and in such games we often encounter equilibria where some information sets are unreached. These equilibria are familiar as the breeding ground for equilibrium refinements, which revolve around specifications of what players can believe and do at unreached information sets. In the strategy space an equilibrium outcome with an unreached information set gives rise to a component of equilibria, with different points in the component corresponding to different specifications of behavior off the equilibrium path. In the state space of a system of differential equations describing evolutionary behavior, this gives rise to a component of stationary states. It is well known, however, that hyperbolic stationary states are isolated. We then cannot dismiss nonhyperbolic stationary states as exceptional cases, no more than we can insist that all Nash equilibria should be strict simply because we can find a space in which such a property is generic. We are forced to consider nonhyperbolic stationary states, with no prospect of rescue by a genericity argument.

The finding that a stationary state is not hyperbolic forces us to question not only the stability properties of the stationary state but also the specification of the evolutionary model that gave rise to such a state. The failure of the stationary state to be structurally stable indicates that important considerations were excluded when constructing the model, out of the mistaken belief that these were sufficiently unimportant to be safely ignored. There are several ways that one might respond, but my preference is to incorporate these neglected factors into the model and then reconsider the stability properties of the stationary states. I refer to this process as "incorporating drift" into the model, with the word "drift" being borrowed from biologists such as Kimura [134].[29]

If we find a stationary state that is not structurally stable, could it be that the phenomenon we are trying to model is simply not structurally stable? It could. Despite the common attempts by economists to force everything into an equilibrium analysis, we have no reason to believe that every evolutionary process is structurally stable. However, no model can establish the conclusion that an underlying process fails to be stable. Instead, there always remains the possibility that the incorporation of currently excluded factors will yield a structurally stable model, and only when we have exactly investigated the underlying process can we conclude that structural stability fails. In the meantime, a finding of instability in a model is an indication that we must look further. Drift must be incorporated into the model to either achieve structural stability or to confirm that the apparent instablity is real.

What should we make of the observation that drift matters? At one level, this appears merely to be the statement that we need better models. Perhaps more importantly, it appears that the outcomes of evolutionary models are quite sensitive to the details of the models. Neither statement seems to be good news for the theory. Much like the refinements literature, there is ample opportunity for the results of evolutionary models to depend upon the models' fine details, all of which

---

29. Kimura popularized the idea of genetic drift, which refers to variations in the genetic composition of a population arising not out of selection pressures but out of the random realizations of the process of reproduction in a finite population. The drift appearing in the models in this book arises from different sources, but shares with Kimura the feature that it has little effect on a population when selection pressures are strong, but potentially great effects when selection pressures are weak.

are likely to be very difficult to observe in applied work. In one respect, even less progress on equilibrium selection is made here than in the refinements literature because evolutionary models introduce a collection of new factors upon which equilibrium selection might depend, including history. However, evolutionary models make progress in two ways. First, the models direct attention to factors that are missed in traditional analyses of games and refinements, and which I suspect play a larger role in shaping equilibria than the rationality considerations found in refinement theories. Second, that drift matters does not mean that we can say nothing about equilibrium, even though we are unlikely to be able to observe the details of drift. Instead, subsequent chapters show that an understanding of how drift matters can be exploited to extract implications for equilibrium selection.

Earlier in this chapter, I compared evolutionary models to the implicit adjustment processes that conventional economic theories use to motivate their focus on equilibrium behavior. The attempt of the tâtonnement literature to provide an adaptive foundation for Walrasian equilibrium is often regarded as a failure because the only general conditions found to be sufficient for convergence were unpalatably strong.[30] There is every indication that evolutionary game theory will also fail to produce strong, generally applicable results. However, I regard this as a virtue rather than a vice. I expect the outcome of a game to depend upon the process by which the players learn to play the game, the specification of the game, and the setting in which the game is played. Only some of these factors are captured by the models of games with which we commonly work, and we accordingly should not expect general results based on these models. Evolutionary game theory provides insight into how outcomes are likely to differ in different classes of games, into what factors outside our models are likely to be important, and into the roles played by these factors.[31]

I have characterized game theory as a language and indicated that the test of the language of game theory is whether it is useful. What does "useful" mean? I take this to mean that without requiring exces-

---

30. See Hahn [110]. For example, the assumptions that the excess demand function satisfies the weak axiom of revealed preference or that all goods are gross substitutes can be used to obtain uniqueness and convergence results, but these assumptions are very demanding.

31. See Mailath [145, p. 263] for similar sentiments.

sively convoluted additional assumptions, such as assumptions about preferences, evolutionary models can produce results describing behavior.[32] Returning to the example set by nonstrategic economic theory, comparative static results describing how behavior can be expected to change as the specification of the game is changed would be especially interesting. The following chapters exploit the nature of drift to make a start on such results. I think this type of work is rendered all the more important by the prospect of using recent developments in experimental economics to evaluate the results.

Equilibrium

Certain basic conclusions concerning equilibrium selection will recur throughout the analysis. The models used will continually lead to Nash equilibria or, in extensive-form games, self-confirming equilibria. This is expected. Attention is focused on the limiting outcomes of dynamic processes that converge. One of the more robust results in evolutionary game theory is the "folk theorem" result that convergence or stability implies Nash equilibrium. However, we shall find little support beyond this for familiar refinements of Nash equilibrium.

The cornerstone of many refinements is the assumption that players will not play weakly dominated strategies, whether this requirement appears explicitly or as an implication of a requirement that players optimize in a world of trembles. Despite the intuitive appeal of a recommendation to avoid weakly dominated strategies, rigorous foundations for such a requirement have been elusive.[33] The evolutionary models below similarly yield outcomes that need not eliminate weakly dominated strategies.

The failure to eliminate weakly dominated strategies is characteristic of a wide variety of evolutionary models. I suspect that if there is any general implication to emerge from evolutionary game theory, it will be that we should be much less eager to discard weakly dominated strategies and should place less emphasis on dominance arguments in

---

32. This is still a subjective definition. I doubt that a more objective definition of "useful" exists, but at the same time have no doubt that subjective criteria of this type lie behind a great deal of the current practice in economics.

33. Examinations of Bayesian rationality, for example, immediately yield the recommendation that one should eliminate strictly dominated strategies, and even iteratively eliminate strictly dominated strategies, but provide much weaker support for weak dominance arguments.

selecting equilibria. This in turn may lead to reconsideration of applied work in a wide variety of areas.

1.6   The Path Ahead

Chapter 2 sets the stage by briefly examining related current work in evolutionary game theory. Chapter 3 introduces a learning model, designed to be the simplest setting in which the relevant points can be made, sacrificing realism for a starker exposition. It then notes that the stationary distribution of the stochastic learning process can be calculated by following the lead of Young [250] and of Kandori, Mailath, and Rob [128], although the behavior of the model over very long periods of time is well approximated (if the population is large) by the deterministic replicator dynamics.

Chapters 4–6 investigate evolutionary models based on deterministic differential equations. Given the results of chapter 3, I interpret this as an analysis of the paths of play generated by an underlying evolutionary process over finite but long periods of time. Chapter 4 establishes conditions ensuring that such dynamics eliminate strictly dominated strategies, but shows that they generally do not eliminate weakly dominated strategies from consideration. Weakly dominated strategies persist even when the dynamics are adjusted so as always to remain in the interior of the state space and hence always to maintain some payoff pressure against dominated strategies. This adjustment is the first appearance of "drift" in the analysis.

The finding that weakly dominated strategies are not eliminated provides the first hint that the outcomes of evolutionary models may bear little resemblance to traditional refinements. Chapter 5 pursues this finding in the context of the Ultimatum Game, a game where the plausibility of equilibrium refinements has been debated. In the Ultimatum Game, player I first makes a proposal as to how a surplus should be split between players I and II. Player II either accepts, implementing the proposed division, or rejects, causing the game to end with payoffs of zero for both players. This game has a unique subgame-perfect equilibrium in which player I demands all of the surplus and player II accepts.[34] In experiments, player I typically offers between $\frac{1}{2}$ and $\frac{1}{3}$ of

---

34. This is the case if the surplus is infinitely divisible. If instead divisions must be made in multiples of a monetary unit, such as a penny, then other subgame-perfect equilibria exist, but player II receives at most a penny in any such equilibrium.

the surplus to player II; offers of $\frac{1}{3}$ or less of the surplus stand at least an equal chance of being rejected. Conventional game theory thus provides a clear equilibrium prediction in a simple game that fares quite badly in experimental settings. In chapter 5, I argue that evolutionary models readily direct attention to Nash equilibria of the Ultimatum Game that are not subgame-perfect and that, like the experimental outcomes, allocate a significant fraction of the surplus to player II.[35]

Chapter 6 examines the forces lying behind the survival of dominated strategies, the implications of these forces for equilibrium selection, and the important role drift plays in developing comparative static predictions.

Chapters 7–9 explore the stationary distributions of learning models that incorporate rare perturbations. It is important to assess how the results of chapters 4–6 hold up over longer periods of time. For example, are weakly dominated strategies eliminated in these stationary distributions? Can we expect to see only subgame-perfect equilibria in the stationary distributions?

Chapter 7 presents the techniques for examining the stationary distribution, based on methods developed by Freidlin and Wentzell [81] and popularized in economics by Foster and Young [79], by Kandori, Mailath, and Rob [128] and by Young [250]. It then conducts a preliminary examination of normal-form games, finding that weakly dominated strategies are in general not eliminated.

Chapter 8 extends this analysis to very simple extensive-form games. The first finding here is that the evolutionary process directs attention not to Nash equilibria but rather to the self-confirming equilibria of Fudenberg and Levine [91]. We then examine self-confirming equilibria satisfying a subgame consistency property of the type that distinguishes subgame-perfect equilibria from Nash equilibria. In the limited class of games examined (extensive-form games of perfect information in which each player moves only once along each path of play), the stationary distribution always contains such an equilibrium. However, there may well be many other outcomes appearing in the stationary distribution. We then have a reason to be interested in the equivalent of a subgame-perfect equilibrium, but not to make that the sole object of our interest.

Chapter 8 shows that the evolutionary model respects forward induction arguments in some simple games. To see why this may be the

---

35. Alternative explanations for the experimental findings are discussed in chapter 5.

case, notice that the motivation for a forward induction argument typically involves a player choosing an action in order to convey information or send a message to another player. The specifics of a forward induction formulation revolve around ensuring that the meaning of the action is unambiguous. In an evolutionary model, the "meaning" attached to an action is determined endogenously by the agents who happen to choose or experiment with the action in the course of the evolutionary process. This meaning will often fluctuate over time and will often not be the one required to drive the forward induction argument. However, it is highly likely that at some point, a meaning gets attached to the message that is sufficient for the forward induction result. If the forward induction equilibrium that appears has sufficiently strong stability properties, the appropriate meaning will be stuck to the message, and the forward induction outcome will persist.

We shall be mostly concerned with the equilibrium refinement issue of evaluating equilibria when players have alternative best replies available. Chapter 9 turns to an examination of the equilibrium selection problem of choosing between strict Nash equilibria. Although it examines a model quite similar to that of Kandori, Mailath, and Rob [128], the chapter assumes that the learning process is noisy, meaning that when players revise their strategies, they sometimes mistakenly switch away from a best reply. Such behavior may be quite likely in the complicated world in which games are actually played. The introduction of such noise has two effects. First, the length of time that must pass before the stationary distribution becomes relevant can be shorter than in models where agents flawlessly switch to best replies. Second, it affects equilibrium selection results, causing the model to select sometimes the payoff-dominant equilibrium and sometimes the risk-dominant equilibrium.

Chapter 10 closes with some thoughts on the additional work required before it is clear whether evolutionary game theory will change the way we think about games or will be a passing fancy. My view here is that the best chance for game theory as a whole to retain its central position in economics is to demonstrate that it is useful in describing how people behave, and that our current best chance of doing this is through evolutionary arguments.

2   The Evolution of Models

This chapter discusses three aspects of evolutionary game theory that play key roles in motivating subsequent chapters. The material in this chapter will be familiar to those working in evolutionary game theory, and is developed in much greater detail in the existing literature on evolutionary games cited at the beginning of chapter 1.

First, although we have much to learn from biological evolutionary models, we must do more than simply borrow techniques from biologists, who have built a theory around the notion of an evolutionarily stable strategy. The difficulty is that evolutionarily stable strategies too often fail to exist. More importantly, they often fail to exist in games that are interesting from an economic point of view and in which one expects evolutionary arguments to have implications.

Second, the response to the existence problem includes a growing list of alternative notions of evolutionary stability, in which many considerations familiar from conventional game-theoretic equilibrium refinements, especially trembles, play a prominent role. The differences in these alternative notions are most readily interpreted as differences in the implicit, dynamic models of the underlying evolutionary process they represent, directing attention to dynamic considerations.

Finally, our best hope for understanding the relationship between evolutionary arguments and equilibrium selection lies in understanding explicitly dynamic evolutionary models, whose literature includes work that borrows heavily from biological models as well as analyses tailored to economic applications. The argument of this book begins with this chapter's observation that two of the most common models can give quite different conclusions for very simple games.

2.1   Evolutionarily Stable Strategies

Symmetric Games

The basic tool for work in evolutionary game theory is the notion of an evolutionarily stable strategy (ESS). The most common setting in which to discuss evolutionary stability is a two-player, symmetric game. Let $S = \{s_1, \ldots, s_n\}$ denote a finite set of pure strategies for a game G. Two players are to play G, each of whom must make a choice from S. Payoffs are specified by the function $\pi : S^2 \to \mathbb{R}$, with $\pi(s_i, s_j)$ being the payoff to either player from using strategy s_i, given that the opponent chooses s_j. Let ΔS denote the set of probability measures on S and interpret an element $\sigma \in \Delta S$ as a mixed strategy. Then we can extend π to $(\Delta S)^2$ in the usual way by letting $\pi(\sigma, \sigma')$ be the payoff to a player choosing mixed strategy σ, given that the opponent chooses σ'.[1]

An evolutionarily stable strategy is defined as follows:

Definition 2.1   Strategy σ^* is an *evolutionarily stable strategy* if there exists $\bar{\epsilon} > 0$ such that for every strategy $\sigma \neq \sigma^*$ with $\sigma \in \Delta S$ and for every $\epsilon < \bar{\epsilon}$,

$$\pi(\sigma^*, \epsilon\sigma + (1 - \epsilon)\sigma^*) > \pi(\sigma, \epsilon\sigma + (1 - \epsilon)\sigma^*).$$

To interpret this definition, we think of this game as being played by a large population of agents. These agents are repeatedly, randomly matched into pairs for a single, static play of the underlying game. Treatments of evolutionarily stable strategies are often somewhat vague as to just how this matching takes place, and there are several points in the story where one might want to see a more precise model. We can think of a sequence of discrete points in time, with every agent being matched with an opponent to play the game once in every period, and with the various configurations in which the agents can be matched being equally likely in any given period.

In each time period, each agent in the population is characterized by a strategy for the game G, either pure or mixed, that the agent plays in that period. Notice that agents are characterized by strategies for G and not for the larger, repeated game created by the matching process.

---

1. Hence, for σ and $\sigma' \in \Delta S$, we have $\pi(\sigma, \sigma') = \sum_{s_i, s_j \in S} \pi(s_i, s_j)\sigma(s_i)\sigma'(s_j)$.

This is typically motivated by some combination of the assumptions that the population is very large, so that the probability of playing a given player twice is so small as to be negligible, and that the matching is anonymous, so that agents are unable to condition their choices on the identity of their current opponent. Current actions then seemingly have no effect on future play, allowing agents to concentrate on strategies of G. When matched to play the game, one of the agents in the pair is randomly assigned to be player I in the game, with the other agent becoming player II and with either assignment being equally likely. Agents do not know which role they fill and hence cannot condition their strategies on the role assignment.

The next step in the story is to assume that there exists some process by which agents choose their strategies and subsequently revise these choices. It suffices to consider a particularly simple case. Suppose that there are just two strategies, σ^* and σ, present in the population, with proportion $1 - \epsilon$ of the population playing σ^*. Given the assumptions that all matches are equally likely and that the population is sufficiently large, we can treat the distribution of strategies in the population and the distribution of strategies among an agent's opponents as being approximately equal.[2] The expected payoff from being matched and playing the game, to an agent playing strategy σ^*, is then $\pi(\sigma^*, \epsilon\sigma + (1 - \epsilon)\sigma^*)$, while the expected payoff to σ is $\pi(\sigma, \epsilon\sigma + (1 - \epsilon)\sigma^*)$.

Now suppose that the strategy revision process causes the proportion of the population playing σ^* to increase if the expected payoff to σ^* exceeds that of σ. In a biological model, the underlying strategy revision process is generally described as one of asexual reproduction, with each agent producing a number of offspring, who play the same strategy, equal to the agent's expected payoff. In an economic context, this process is one in which agents learn that some strategies yield higher payoffs than others and switch to the former. In either case, if σ^* is an ESS, then by definition σ^* earns a higher payoff than any other strategy σ whenever most of the population plays σ^*. The appearance of a small group of agents playing strategy σ in a population of agents playing the ESS strategy σ^*, whether arising from mutations in a biological model or experimentation in an economic model, will then give rise to adjustments that increase the proportion of the population

---

2. Boylan [49] and Gilboa and Matsui [100] examine the technical issues involved in constructing a model of large populations with equally likely matches.

playing strategy σ^*, eventually eliminating σ and restoring a state in which all agents play σ^*. The strategy σ^* can thus repulse a sufficiently small invasion of strategy σ.

As its name suggests, the concept of an ESS is a stability condition. The interpretation of the concept begins with a state in which all of the population plays strategy σ^*. No comment is made as to how such a state has been reached. In addition, the strategy σ^* is asked to be stable only in the face of isolated bursts of mutations. The model implicitly allows a mutant σ to be purged from the population before the appearance of the next mutant. This would be an especially serious limitation if mutants were required to play pure strategies. It appears less demanding when we recognize that, by playing a mixed strategy, a mutant can capture some of the effect of the simultaneous appearance of mutants playing a number of different strategies.

The significance of the evolutionarily stable strategy concept is thus that if nature (in a biological model) or some process of learning (in an economic model) has a tendency to produce monomorphic population states, and if occasional mutations also occur, then the states we should expect to observe are ESSs, which can persist in the face of such mutations.

The bilinearity of the payoff function can be used to derive the following equivalent and commonly used definition of an ESS:

Definition 2.2 Strategy $\sigma^* \in \Delta S$ is an *evolutionarily stable strategy* if, for all $\sigma \neq \sigma^*$:

$$\pi(\sigma^*, \sigma^*) \geq \pi(\sigma, \sigma^*) \tag{2.1}$$

$$\pi(\sigma^*, \sigma^*) = \pi(\sigma, \sigma^*) \Rightarrow \pi(\sigma^*, \sigma) > \pi(\sigma, \sigma). \tag{2.2}$$

This, the original, definition of an ESS offered by Maynard Smith [149] and by Maynard Smith and Price [150] has the advantage of making it clear that the ESS condition is the combination of a Nash equilibrium requirement, given by (2.1), and a stability requirement, given by (2.2). The latter ensures that the ESS σ^* can repel mutants. In order to vanquish a mutant, the strategy σ^* must fare better than a mutant σ in a population almost entirely composed of σ^*s but with a few σs, where "few" can be made as small as we would like. Because the population is almost all σ^*s, this requires that σ^* be a best reply against itself. It suffices that σ^* be a strict best reply to itself. If an alternative best reply σ exists, then σ^* must be a better reply to σ than is σ itself. If conditions (2.1)–(2.2) fail, then the strategy σ is said to be able to invade σ^*.

Asymmetric Games

In many cases, we will be interested in asymmetric games. In biological applications, asymmetries may arise because two animals in a game are of different sizes, or have identities as the owner of a territory and the intruder into that territory. In economic applications, asymmetries will often arise because the agents in a transaction have identities such as buyer and seller, firm and union, principal and agent, employed and unemployed, or informed and uninformed.

There are two ways an asymmetry can appear. First, the players in a game may have different strategy sets or different payoff functions. In a biological contest, the game may be played by different species, one of whom might be predator and one of whom might be prey. In an economic context, we might be dealing with firms and workers. Second, the players may have identical strategy sets and payoff functions, but may know whether they are player I or II and may be able to condition their strategies on this information. For example, the labels "player I" and "player II" may correspond to characteristics such as age, size, height, or occupation. Even if these characteristics have no effect on payoffs, they can potentially be used by players to create asymmetries on which they can condition their strategies. This seemingly minor difference can have important implications. When discussing the coordination game of figure 1.4, for example, I assumed that the game was asymmetric, so that (T, R) was a pure Nash equilibrium, even though the players have identical strategy sets and payoff functions. If this were a symmetric game (with strategies T and L perhaps relabeled as A, and strategies B and R relabeled as B), then there is a unique Nash equilibrium (and ESS) in which each player mixes equally between the two strategies.

In order to apply the ESS concept to asymmetric games, we now think of agents, when called upon to play the game, being randomly (with equal probabilities) assigned to be either player I or player II. The agents are informed of the role to which they have been assigned and are allowed to condition their choice upon this assignment. Hence, the underlying asymmetric game (denoted by G) has been embedded in a larger symmetric game (denoted by Γ). The first move in the latter game is Nature's, who decides which of the two agents is player I and which is player II. A strategy in game Γ is then a choice of a strategy to be played in the role of player I in game G and a strategy to be played in the role of player II in game G. Payoffs in Γ are given by the expectation over the payoffs attained as each of the players in game G. We then apply the standard ESS condition to game Γ.[3]

While this is the most common method for applying ideas of evolutionary stability to asymmetric games, it is clearly inappropriate in some cases. If the asymmetry arises out of the agents being members of different species, then it is unrealistic to think of a given agent sometimes filling the role of player I and sometimes filling the role of player II. If the asymmetry involves buyers and sellers, and the good in question is a financial asset or a used car, then it may well be realistic to model an agent as sometimes buying and sometimes selling. If the good is a new car or a service such as extinguishing oil well fires, then it is probably more appropriate to model agents as always being on only one side of the transaction. If the asymmetries arise out of the ability of agents to condition their strategies on factors such as age, size, height, or occupation, with conventions such as always choosing the older agent in a pair to be player II, then it may be unrealistic to model all agents as being equally likely to fill either role.[4]

Equilibrium

Biologists have made frequent use of models based on evolutionarily stable strategies, in some cases finding the models to be remarkably good tools for describing behavior (cf. Hammerstein and Riechert [112] and Maynard Smith [149]). Much of the interest in evolutionary game theory in economics arises out of the possible connections between evolutionary theories and more traditional game-theoretic solution concepts. What are the equilibrium properties of an ESS?

It will be convenient to consider only symmetric games. A connection between evolutionarily stable strategies and Nash equilibrium has already been noted:[5]

---

3. This extends the ESS concept to asymmetric normal-form games. The question remains of how to apply the concept of extensive-form games. One possibility is to simply work with the corresponding normal form. Selten [211, 212] presents an alternative approach that explicitly exploits the extensive-form structure of the game.
4. A notion of evolutionary stability for the case of an asymmetry arising because the game is played by two distinct groups of agents, such as predator and prey or buyers and sellers, is given by Cressman [68]. A more common approach to asymmetric games in economics is to work with dynamic models. This chapter touches on dynamic models in section 2.3.
5. Because the results in this chapter are well known, I do not present proofs of its propositions.

Proposition 2.1 If σ^* is an evolutionarily stable strategy of game G, then (σ^*, σ^*) is a Nash equilibrium of G that is isolated in the set of symmetric Nash equilibria. If (σ^*, σ^*) is a strict Nash equilibrium, then σ^* is an evolutionarily stable strategy.

That every ESS is a Nash equilibrium follows immediately from condition (2.1) of definition 2.2, which requires that the evolutionarily stable strategy be a best reply to itself. To see that a strict Nash equilibrium must be an ESS, we note that condition (2.2) in definition 2.2 comes into play only in the case of alternative best replies. Because a strict Nash equilibrium has no alternative best replies, it then must be an ESS. Hofbauer and Sigmund [119, p. 121] show that an ESS σ^* must be contained in a neighborhood with the property that σ^* is a better reply to σ than is σ itself, for any other strategy σ in the neighborhood. This leads to the conclusion that an ESS must be an isolated Nash equilibrium.

Definition 2.2 makes it clear that evolutionary stability is a stronger equilibrium concept than Nash equilibrium. An indication of how much stronger is the ESS concept is provided by the following, taken from van Damme [240, theorem 9.3.4]:

Proposition 2.2 If σ^* is an evolutionarily stable strategy, then (σ^*, σ^*) is a proper equilibrium.

It is useful to sketch the intuition behind this result, for a particularly transparent special case, because it provides the first indication of how refinement ideas can appear in an evolutionary context. Let s_1 be a pure strategy that is an ESS. If (s_1, s_1) is a strict Nash equilibrium, then it is certainly proper. What if there are alternative best replies? Let the strategy set be denoted by $S = \{s_1, s_2, \ldots, s_n\}$ with the numbering chosen so that

$$\pi(s_1, s_1) = \pi(s_2, s_1) \tag{2.3}$$

$$\pi(s_i, s_1) < \pi(s_{i-1}, s_1), \quad 3 \leq i \leq n. \tag{2.4}$$

Hence, in this special case there is just one alternative best reply (namely, s_2) and no other payoff ties against s_1. This is the simplest context in which a pure-strategy equilibrium might fail to be proper.

Is (s_1, s_1) a proper equilibrium? This is essentially the question of whether choosing s_1 is the right way to break the payoff tie between strategies s_1 and s_2. From the standpoint of proper equilibrium, this tie has been broken correctly if s_1 continues to be a best reply when an appropriate type of tremble is introduced into the game. More precisely, (s_1, s_1) is proper (Myerson [160]) if we can find a sequence σ_n, where each σ_n is a mixed strategy with full support on S and where $\sigma_n(s_1) \to 1$, and if we can also find a corresponding sequence ϵ_n with $\epsilon_n \to 0$ and with

$$\pi(s_i, \sigma_n) < \pi(s_j, \sigma_n) \Rightarrow \sigma_n(s_i) \leq \epsilon_n \sigma_n(s_j) \tag{2.5}$$

for all $s_i, s_j \in S$. The motivation behind these trembles is that players should be less likely to make small mistakes than big ones. Condition (2.5) is the statement that the probability with which players tremble to low-payoff strategies must be much smaller than the probability with which they tremble to higher-payoff strategies.

Conditions (2.3)–(2.4) establish a payoff ranking among strategies, and the obvious approach to showing that (s_1, s_1) is proper is to mimic this ranking in the specification of trembles. Fix a sequence ϵ_n with $\epsilon_n \to 0$ and, as a candidate for σ_n, let $\sigma_n(s_i) = (\epsilon_n)^i$ for $i \geq 2$, with $\sigma_n(s_1)$ comprising the remaining probability. This ensures that low-payoff strategies receive smaller tremble probabilities; from (2.4), it is then immediate that (2.5) holds for all sufficiently large n and for all strategies, except possibly the case of s_1 and s_2. That is, the key issue in determining whether (s_1, s_1) is proper involves the alternative best reply s_2 and not the other, inferior replies.

We are then left with breaking the tie between s_1 and s_2, which means we must establish that $\pi(s_1, \sigma_n) \geq \pi(s_2, \sigma_n)$. Given that the trembles make s_1 the most likely strategy, the desired inequality will certainly hold if $\pi(s_1, s_1) > \pi(s_2, s_1)$. However, the assumption (2.3) that s_2 is an alternative best reply gives equality here. We accordingly appeal to the next most likely strategy, which is s_2, and it suffices that $\pi(s_1, s_2) > \pi(s_2, s_2)$. But this inequality must hold because s_1 is an ESS, and condition (2.2) then ensures that $\pi(s_1, s_2) > \pi(s_2, s_2)$. Hence (s_1, s_1) is proper.

The key observation here is that both the ESS concept and properness subject s_1 to a lexicographic test. It suffices in each case that s_1 be a strict best reply to itself. If it is only a weak best reply because s_2 is also a best reply, then properness asks us to regard s_2 as the next most likely strategy, and it suffices that s_1 be a better reply to this next most likely strategy of s_2 than is s_2. But the ESS concept also requires s_1 to be a better reply to s_2 that is s_2 itself, as does the ESS concept for each alternative strategy, and hence ESS suffices for properness.[6] More generally, an equilibrium refinement requires an equilibrium to survive trembles that inject out-of-equilibrium strategies into the game. The notion of an evolutionarily stable strategy analogously requires a strategy to survive mutations that inject unplayed strategies into the game. More sophisticated refinements place restrictions on the trembles, such as the properness requirement that players are less likely to tremble to lower-payoff strategies. If the evolutionary process is more effective at quickly eliminating low-payoff strategies, then we would similarly expect relatively high-payoff mutations to have more effect than lower-payoff mutations. It is then no surprise that many refinement ideas should reappear in an evolutionary setting.

Van Damme [237] and Kohlberg and Mertens [135] have shown that a proper equilibrium induces a sequential equilibrium in every corresponding extensive form. An evolutionarily stable strategy thus has strong refinement properties, yielding a proper equilibrium in the normal form and sequential equilibrium in the extensive form, and does so without the tedium of finding sequences of trembles.

Existence

The ESS concept clearly leads us back to equilibrium concepts that have occupied center stage in the refinements literature. However, another characteristic of evolutionarily stable strategies has limited their use in studying equilibrium issues: an ESS often fails to exist.

The existence of an evolutionarily stable strategy poses no difficulties in 2×2 symmetric games. van Damme [240, theorem 9.2.3] shows

Proposition 2.3 Let G be a two-player symmetric game with $|S| = 2$ and with $\pi(s_1, s_1) \neq \pi(s_2, s_1)$ and $\pi(s_1, s_2) \neq \pi(s_2, s_2)$. Then G has an evolutionarily stable strategy.

To motivate the payoff inequalities that are assumed in this proposition, consider the most extreme case where they fail, namely, where all of the payoffs in the game are the same. In this case there is no hope for an inequality in either of (2.1)–(2.2), and hence no hope for an ESS.

---

6. In the simple case just considered, the properness story stops with the consideration of s_2 because of three factors: the ESS is a pure strategy, there is only one alternative best reply, and no two inferior best replies receive equal payoffs against the ESS. Relaxing any of these introduces some payoff ties that must be handled carefully when constructing the sequence of trembles σ_n. Doing so introduces some new technical issues but no new conceptual ones.

	H	T
H	1, −1	−1, 1
T	−1, 1	1, −1

Figure 2.1
Matching Pennies

	s_1	s_2	s_3
s_1	1, 1	2, −2	−2, 2
s_2	−2, 2	1, 1	2, −2
s_3	2, −2	−2, 2	1, 1

Figure 2.2
A game with no ESS

Considerable progress has been be made in economic and biological applications by working with two-player 2×2 games. A great deal of economics can be done within the metaphors provided by the Prisoners' Dilemma, coordination games, and games with only mixed-strategy equilibria such as Chicken and Matching Pennies (the last shown in figure 2.1).[7] However, these are not the only important games, and there is unfortunately little scope for generalizing this existence result beyond such games. Van Damme [240, figure 9.2.1] presents an example of a 3×3 game with no ESS. A slightly modified version of this example is provided by figure 2.2. This game has a unique Nash equilibrium given by $\sigma^* = (\frac{1}{3}, \frac{1}{3}, \frac{1}{3})$ for payoffs of (1, 1). Every strategy is a best reply to σ^*. In addition, strategy s_1 (or any other pure strategy) satisfies $\pi(s_1, s_1) = \pi(\sigma^*, s_1) = 1$, ensuring that σ^* is not an ESS.

The nonexistence of an ESS in figure 2.2 brings us back to a familiar and recurring point. We often find it easy to motivate a strict Nash equilibrium, where agents have unique payoff-maximizing choices. Upon first encountering a Nash equilibrium that is not strict, students often ask how the agents happen to choose precisely the right strategy from the set of alternative best replies. We have no better response than "because that is what's required for an equilibrium." This is commonly

---

7. Figures 1.3 and 1.5 contain a coordination game and the Prisoner's Dilemma; figure 3.4 contains Chicken, and figure 2.1 contains Matching Pennies.

followed by some dynamic intuition about how other choices would prompt adjustments in strategies until, one hopes, an equilibrium is reached.

The ESS concept, and its potential nonexistence, tells us something about this dynamic intuition. Inspecting figure 1.4 shows that if we consider this to be a symmetric game (with strategies labeled $T = L = A$ and $B = R = B$), then injecting a nonequilibrium strategy, such as playing A with probability $\frac{2}{3}$, into a population playing the equilibrium half-half mixture, causes the former to receive a strictly lower pay-off. This suggests that some pressure will be brought to bear against nonequilibrium strategies in this case, and the equal mixture is an ESS as well as a Nash equilibrium.[8] There may then be some basis for the dynamic intuition behind the selection of the best replies required to support the Nash equilibrium. In figure 2.2, however, the introduction of a nonequilibrium strategy does not introduce a similar payoff pressure. This challenges the intuition offered in support of the Nash equilibrium in this case, and precludes the existence of an ESS.

It may not appear particularly troubling that the game in figure 2.2 has no ESS. The intuitive stories offered as a defense of Nash equilibrium in the presence of alternative best replies may be less convincing in this case, and it may not be clear how an evolutionary process should be expected to behave on this game. However, the ESS concept encounters more serious existence difficulties. An ESS can fail to exist even though the Nash equilibrium concept is convincing, and even though the implications of an evolutionary process appear to be intuitively obvious. Consider figure 2.3. There is a single component of Nash equilibria in this game, including any mixture over s_1 and s_2. The unique Nash equilibrium in undominated strategies is (s_1, s_1). There is no ESS. In particular, s_1 fails to be an ESS because s_2 gives equality in both of (2.1)–(2.2). However, strategy s_3 always earns strictly less than any other strategy in the population, and it seems clear that an evolutionary model will at least agree with the Nash equilibrium concept in yielding the prediction that strategy s_3 will be eliminated. The ESS concept can thus fail to exist in cases where evolutionary arguments have implications we would like our model to capture.

8. Notice that the same is not true of Matching Pennies (figure 2.1) because this is not a symmetric game. An ESS must then be a strict Nash equilibrium (see proposition 2.4 below), which does not exist for this game.

	s_1	s_2	s_3
s_1	1, 1	1, 1	1, 1
s_2	1, 1	1, 1	0, 0
s_3	1, 1	0, 0	0, 0

Figure 2.3
A game with no ESS

The games in figures 2.2 and 2.3 are not generic, where the word "generic" is given its commonly used, informal meaning of involving no payoff ties. I am sympathetic to the contention that game theory should be confined to generic games. The models with which we work are approximations, including the numbers we choose to represent payoffs. We may find it convenient in constructing the model to assign the same payoff to two seemingly similar outcomes. But if this payoff tie turns out to play a major role in driving the results, then we must reconsider our modeling assumptions. It is unlikely the two outcomes have literally identical payoffs; that the relative magnitude of their payoffs matters is a signal we should calculate these payoffs more carefully and adjust our model accordingly. We should then be suspicious of any results driven by a nongenericity in the specification of the game.

At the same time, care must be taken in formulating a genericity requirement. It is well known that we could dispense with most equilibrium refinement considerations entirely, simply by restricting attention to games that are "generic" in the sense that there are no ties in their *normal-form* payoff matrices. This observation has failed to forestall work on equilibrium refinements because the genericity requirement is too strong. Normal-form payoff ties can appear because two normal-form strategy profiles lead to the same terminal node in a generic extensive-form game. We cannot avoid a tie in this case and cannot raise genericity-based objections to such a game. Even asking for no ties to appear in extensive-form payoffs can be too strong. The appropriate place in which to require genericity is not necessarily the space of payoffs, in either the normal or extensive forms, but rather the mapping from economic outcomes into utilities. This mapping may be generic, and yet we may have to contend with ties in payoffs. These ties may appear in both the normal and extensive form because different strategies give rise to the same economic outcome. For example, I consider Rubinstein's alternating offers bargaining game [197] to be generic even though it generates ties in the space of extensive-form payoffs. The existence of an ESS then cannot be rescued by a deftly chosen genericity assumption.

The existence problems with evolutionarily stable strategies become more dramatic when examining asymmetric games. Recall that if G is an asymmetric game, then a strategy σ^* in its symmetric version Γ specifies what to do in game G in the roles of both players I and II. Write this as $\sigma^* = (\sigma_I^*, \sigma_{II}^*)$. The following result is due to Selten [210]:

Proposition 2.4   Let G be an asymmetric normal-form game and let Γ be the symmetric version of G. Then σ^* is an evolutionarily stable strategy of Γ if and only if $(\sigma_I^*, \sigma_{II}^*)$ is a strict Nash equilibrium of G.

The intuition behind this result is straightforward. Let σ^* be a strategy in Γ, so that σ^* corresponds to a strategy pair $(\sigma_I^*, \sigma_{II}^*)$ in G. Suppose that $(\sigma_I^*, \sigma_{II}^*)$ is a Nash equilibrium but not a strict Nash equilibrium of G. Then an alternative best reply must exist. Let σ_I' be an alternative best reply to σ_{II}^* in game G. Now let σ' be the strategy in game Γ given by $(\sigma_I', \sigma_{II}^*)$, so that σ' plays σ_I' in role I and σ_{II}^* in role II. In order for σ^* to be an ESS, σ^* must be able to repulse small invasions of σ', meaning that the expected payoff to σ^* must be higher than the expected payoff to σ' once the latter has appeared in the population. In this case, however, σ^* and σ' will have identical expected payoffs. Because σ_I^* and σ_I' are alternative best replies to σ_{II}^*, strategies σ^* and σ' must secure the same expected payoffs when they are called upon to play in role I. In role II, σ^* and σ' feature identical behavior, and again must secure the same expected payoffs. There is then no selection pressure against strategy σ', and σ^* is therefore not an ESS.

The asymmetry of the game G plays a crucial role in this result. In symmetric games evolutionarily stable strategies exist to which there are alternative best replies. An example is the half-half mixture in the symmetric version of figure 1.4. An alternative best reply to such an ESS does as well against the ESS as does the latter, but the alternative best reply sometimes meets itself and then does worse than the ESS. The key to such a result is that when the alternative best reply meets itself, behavior in *both* roles of the game is altered. In a symmetric game, where strategies cannot be conditioned on roles, a mutant cannot help but affect behavior in both roles. When G is asymmetric, however, mutants such as σ' can appear that affect behavior in only one role of the game. As long as this altered behavior still involves a best reply, the mutant remains a best response to the candidate ESS. The same payoffs appear when two mutants meet as when a mutant and the candidate ESS meet, ensuring that the latter is not an ESS. A candidate for an ESS can then be sure of repelling all mutants only if no mutant is a best reply, implying that an ESS must be a strict Nash equilibrium.

Restricting attention to strict Nash equilibria is a sufficiently strong requirement to severely limit the usefulness of the ESS concept in asymmetric games. Mixed strategies, such as the unique Nash equilibrium of Matching Pennies (figure 2.1), are immediately precluded. In extensive-form games this requires that equilibrium strategies be pure and reach every information set. However, interesting economic problems readily lead to games and equilibria in which some information sets are not reached. Our intuition is that game theory in general and evolutionary game theory in particular should have something to say about such games. Making progress on refining and selecting equilibria then requires some alternative to working with the notion of an evolutionarily stable strategy.

It is important to note what is *not* being asserted in the previous paragraph. It is not the case that I have already selected a favorite equilibrium concept and am now disappointed because evolutionary stability, in insisting on strict Nash equilibria, fails to agree with that concept. To the contrary, I think there is little to be gained from seeking a single equilibrium concept to be ordained on abstract grounds as the "right" one. Instead, the appropriate equilibrium is likely to vary across games and settings, and may often depend upon institutional considerations that are not captured in the specification of the game. As a result, we are often faced with the task of interpreting and evaluating equilibria, and these equilibria are typically not strict Nash equilibria. We need some help in evaluating these equilibria, and this is what I seek from an evolutionary model. A model that cannot help in this task is disappointing not because if fails to select the right equilibrium but because it fails to provide information that might help identify some equilibria as more interesting than others.

2.2 Alternative Notions of Evolutionary Stability

The previous section contains good news and bad news. The concept of evolutionary stability captures considerations familiar from the refinements literature, but does so in a way that all too easily leads to nonexistence. One response to the potential nonexistence of evolutionarily stable strategies has been to propose alternative notions of evo-

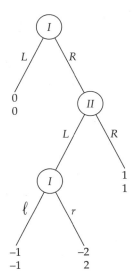

Figure 2.4
Game G^*

lutionary stability.[9] Many of the ideas involved in these alternatives, including the use of trembles, are again familiar from the refinements-of-Nash-equilibrium literature.

The list of alternative evolutionary stability notions is long and growing. Rather than attempt a catalog of such notions, I concentrate on a few alternatives that illustrate two points. The first is that trembles play a key role in these equilibrium concepts, while the second is that differences in the alternative equilibrium concepts can be traced to differences in the implicitly dynamic models that lie behind them.

To focus the discussion, it will be helpful to keep a running example. Consider the extensive-form game G^* shown in figure 2.4. This perfect information game has three pure-strategy Nash equilibria, given by $(L\ell, L)$, (Lr, L) and $(R\ell, R)$. The first two are contained in a single component in which player I plays any mixture of $L\ell$ and Lr and player II plays L with probability at least $\frac{1}{2}$. The third is contained in a component in which II plays R and player I mixes between $R\ell$ and Rr, with probability at least $\frac{1}{3}$ on $R\ell$. There is a unique subgame-perfect equilibrium, given by $(R\ell, R)$. What equilibrium, if any, will be selected by an evolutionary process?

_____

9. The material in section 2.2 is taken from Binmore and Samuelson [25].

We can first ask about evolutionarily stable strategies for game G^*. This requires working with the symmetrized version Γ^* of game G^*, in which a strategy for an agent includes a strategy for both players I and II and an agent is equally likely to fill the role of either player I or player II. Because an ESS of Γ^* must be a strict Nash equilibrium of G^* and the latter has no strict Nash equilibria, we immediately have[10]

Remark 2.1 An evolutionarily stable strategy does not exist for Game G^*.

For example, the subgame-perfect equilibrium $(R\ell, R)$ is not evolutionarily stable because the strategy (Rr, R) can invade. More precisely, taking σ^* to be $(R\ell, R)$ and σ to be (Rr, R) yields equality throughout conditions (2.1)–(2.2). The strategy (Rr, R) can invade because it differs from the subgame-perfect equilibrium strategy $(R\ell, R)$ only at information sets that are never reached when $(R\ell, R)$ and (Rr, R) play each other or themselves. The different behavior specified by the mutant (Rr, R) then has no effect on play. There is no obvious way for an evolutionary process to bring pressure against this strategy, and the subgame-perfect equilibrium cannot be evolutionarily stable.

Neutrally Stable Strategies

The strategies $(R\ell, R)$ and (Rr, R) give identical outcomes in game G^*. Then why do we care whether (Rr, R) can invade? More generally, we may not want to discard an equilibrium, even if it is not evolutionarily stable, as long as the only invaders are those whose behavior and payoffs match that of the candidate strategy. Why worry about changes in behavior off the equilibrium path? Toward this end, Maynard Smith [149, p. 107] defines the concept of a neutrally stable strategy. Let S be the set of pure strategies in an asymmetric game G, so that $\Delta S \times \Delta S$ is the set of mixed strategies for the symmetrized game Γ.[11] Then we have[12]

---

10. It is common to say that G^* has no ESS, though the asymmetry of G^* forces the ESS concept to be applied to the symmetrized version of the game given by Γ^*. Analogous references to G^* are made for subsequent equilibrium concepts.

11. As usual, a mixed strategy for Γ is taken to be two independent mixtures over S, one for each player in game G.

12. Because G^* is asymmetric, I state definitions in this section in terms of strategies for asymmetric games. The usual symmetrization is involved in computing payoffs. The profit functions that appear in this definition are then the expected profits, with the expectation taken over the identity of the opponent and the role to be occupied when playing the game.

Definition 2.3   A *neutrally stable strategy* for an asymmetric game G is a strategy $\sigma^* \in \Delta S \times \Delta S$ such that, for all $\sigma \in \Delta S \times \Delta S$:

$$\pi(\sigma^*,\sigma^*) \geq \pi(\sigma,\sigma^*) \tag{2.6}$$

$$\pi(\sigma^*,\sigma^*) = \pi(\sigma,\sigma^*) \Rightarrow \pi(\sigma^*,\sigma) \geq \pi(\sigma,\sigma). \tag{2.7}$$

Comparing these conditions with (2.1)–(2.2), we see that the final strict inequality in (2.2) has been replaced with a weak inequality. The effect of this change is that we do not require a neutrally stable strategy to receive a strictly higher payoff than any mutant, and hence cannot be sure that every mutant will be driven out once it appears. At the same time, (2.6)–(2.7) ensure that a mutant cannot earn a higher payoff than the neutrally stable strategy, and hence that the mutant cannot displace the neutrally stable strategy. It thus may be that after a mutant appears, the mutant and the neutrally stable strategy coexist.

It is straightforward to verify that neutrally stable strategies exist for game G^*:

Remark 2.2   In Game G^*, every Nash equilibrium (σ_I, σ_{II}) is a neutrally stable strategy except (i) the Nash equilibrium in which player II plays R and player I plays $R\ell$ with probability $\frac{1}{3}$ and Rr with probability $\frac{2}{3}$; and (ii) any Nash equilibrium in which player I mixes between $L\ell$ and Lr, while player II plays L with probability $\frac{1}{2}$.

Every Nash equilibrium is thus neutrally stable except those teetering just on the brink of not being Nash equilibria.

The difficulty with neutrally stable strategies is that they may not be "stable enough." Consider the subgame-perfect equilibrium of Game G^*. How reassuring do we find it that this strategy is neutrally stable? Beginning in a state in which all agents in the population play their part of the subgame-perfect equilibrium, changes in behavior can occur that do not give rise to any evolutionary pressure to adjust strategies. In particular, mutations can introduce the strategy (Rr, R). An outcome in which $(R\ell, R)$ is played can then be displaced by one in which player II plays R and player I mixes between $R\ell$ and Rr, as long as suitably large probability (i.e., at least $\frac{1}{3}$) is still placed on $R\ell$. This shift has no implications for observed behavior. However, this may lead to a population polymorphism in which $\frac{2}{3}$ of the agents in population I play Rr. At this point, small infusions of mutants playing Rr and L will cause L to receive a higher payoff for player II than R, causing strategy L to grow and potentially disrupting the subgame-perfect equilibrium outcome. Neutral stability may then be too weak a requirement to be interesting. At the very least, it seems that when working with neutral stability, one should address polymorphic populations and should assess the stability of entire components of equilibria.

Trembles

The objection to the neutral stability of the subgame-perfect equilibrium in the previous paragraph is based on the *possibility* that mutations can lead the system to an outcome in which Rr is played by a proportion of the population approaching $\frac{2}{3}$. How probable are such mutations? If the mutations needed to cause stability problems for a neutrally stable strategy are quite unlikely, then the neutrally stable strategy may not be problematic. Alternatively, we might accept a component of neutrally stable strategies as stable as long as mutations will direct the system to the "more stable" part of the component.

Maynard Smith [149] suggests that small perturbations in either the game or the evolutionary process will eliminate the effect of mutations on off-the-equilibrium-path behavior. In particular, perturbations in Game G^* might take the form of some player-II agents occasionally playing L, even though the strategy of each such agent is to play R. These perturbations would cause player I agents to earn a higher payoff from playing $R\ell$ than from Rr and hence should push player I back to $R\ell$, interrupting any movement toward Rr.

Selten [211, 212] pursues a similar intuition in introducing the concept of a *limit evolutionarily stable strategy*.[13] To find a limit ESS of an asymmetric game G, we first construct a perturbed game $G(\epsilon)$, where $\epsilon = (\epsilon_I, \epsilon_{II})$ is a vector containing a nonnegative number for each of the pure strategies for each of players I and II. The players and payoff functions are unchanged, but players in $G(\epsilon)$ must choose mixed strategies from the sets $\Delta_I(\epsilon_I)$ and $\Delta_{II}(\epsilon_{II})$, where $\sigma \in \Delta_I(\epsilon_I) \Rightarrow \sigma(s_1) \geq \epsilon_I(s_1) \geq 0$ for any pure strategy $s_1 \in S_I$, with a similar characterization of Δ_{II}. Hence players can be forced to attach positive probability to certain pure strategies in the perturbed game. Notice, however, that $\epsilon_I(s_1) > 0$ is not required, so that strategies in the perturbed game need

---

13. The limit ESS concept is Selten's first step in the construction of evolutionary stability ideas that explicitly take account of the extensive-form structure of a game [211, 212]. Elementary extensive-form considerations reappear in chapter 8 below.

not be completely mixed, unlike the case of a perfect equilibrium. The game G itself is thus always also a perturbed game. This has the immediate consequence that any evolutionarily stable strategy is also a limit evolutionarily stable strategy.

The next step is to construct a sequence of perturbed games converging to G. An ESS for each of these perturbed games is selected, trusting the perturbations to ensure that these ESSs exist. Our interest is then directed to the limit of these ESSs as the perturbations disappear, which is a limit ESS:

Definition 2.4   The strategy profile σ^* is a *limit evolutionarily stable strategy* of the asymmetric normal form G if there is a sequence of perturbed games $G(\epsilon(n))$ and evolutionarily stable strategies $\sigma(n)$ of $G(\epsilon(n))$ with the properties that $\lim_{n \to \infty} \epsilon(n) = 0$, and $\lim_{n \to \infty} \sigma(n) = \sigma^*$.

Samuelson [199] shows that a strategy profile is a limit ESS if and only if it is a pure-strategy perfect equilibrium that does not attach positive probability to any strategy for a player i with the property that another strategy for player i yields identical payoffs for any opponent strategy.[14]

Hofbauer and Sigmund [119, p. 288] offer an alternative method for incorporating the effect of trembles, namely, the concept of a *weak evolutionarily stable strategy* for asymmetric normal-form games:[15]

Definition 2.5   Let S_I and S_{II} be players I and II's pure strategy sets in an asymmetric game. Then $(s_I^*, s_{II}^*) \in S_I \times S_{II}$ is a *weak evolutionarily stable strategy* if it is a Nash equilibrium and for $i \in \{I, II\}$ and $j \neq i$,

$$\pi_i(s_i^*, s_j^*) = \pi_i(s_i, s_j^*) \Rightarrow \pi_i(s_i^*, s_j) > \pi_i(s_i, s_j) \; \forall s_j \in S_j \setminus \{s_j^*\}. \tag{2.8}$$

Hence alternative best replies for player i against player j's equilibrium strategy must fare strictly worse against all other strategies for j.

---

14. More formally, positive probability must not be attached to s_i if these exists s_i' such that $\pi_i(s_i, s_{-i}) = \pi_i(s_i', s_{-i})$ for all s_{-i}. Notice that this is not simply the requirement that the game be in reduced normal form because only i's payoffs are involved in the condition.

15. It is immediate that a mixed strategy cannot be weakly evolutionarily stable. To see this, let $(\sigma_I^*, \sigma_{II}^*)$ be a weak evolutionarily stable strategy with σ_I^* mixed. Then let $\sigma_{II} \neq \sigma_{II}^*$. Because every pure strategy in the support of σ_I^* is an alternative best reply to σ_{II}^*, the definition of weak evolutionary stability requires that every pure strategy in the support of σ_I^* earn a payoff against σ_{II} that is strictly inferior to the payoff of σ_I^* against σ_{II}, which is impossible.

To see the motivation for this concept, suppose that (s_I^*, s_{II}^*) is a weak ESS and suppose that player II's strategies are subject to trembles. No matter what the nature of these trembles, as long as they are not too large, condition (2.8) ensures that strategy s_I^* earns a strictly higher payoff for player I than any other strategy. This produces a force pushing player I back to the equilibrium strategy s_I^*.

It is easily verified that if we simply apply the concept of a limit evolutionarily stable strategy or weak evolutionarily stable strategy to G^*, we find[16]

Remark 2.3 Game G^* has a unique limit evolutionarily stable strategy and weak evolutionarily stable strategy, given by the subgame perfect equilibrium of $(R\ell, R)$.

The concepts of limit and weak evolutionary stability provide some insight into the role of trembles in evaluating equilibria. In the refinements literature trembles are used to eliminate Nash equilibria, while in evolutionary models trembles are used to *create* evolutionarily stable strategies. To see how this difference arises, notice that trembles can have the effect of breaking payoff ties between alternative best replies. Nash equilibria readily accommodate alternative best replies, even ones we might not wish to accommodate, such as cases in which an equilibrium assigns probability to a strategy that is weakly dominated by an alternative best reply. Trembles eliminate such equilibria by converting some of the alternative best replies (such as the dominating strategy) into superior replies, and then requiring that no probability be assigned to strategies whose payoffs lag behind in the presence of trembles.

In contrast, we have seen that the existence of alternative best replies often has the effect of preventing a strategy from being an ESS. Trembles once again can break the ties between the various alternative best replies. If σ^* is the strategy of interest, then one possibility is that the trembles turn alternative best replies into inferior replies. If this is the case, then σ^* is a good candidate for satisfying a tremble-based evolutionary stability notion.

---

16. The other two candidates for a limit evolutionarily stable strategy, (Lr, L) and $(L\ell, L)$, satisfy the condition of being pure-strategy trembling-hand-perfect equilibria, but the fact that Lr and $L\ell$ produce the same payoffs to player I regardless of player II's action ensures that these are not limit ESSs. This may suggest that we should apply the limit ESS concept to the reduced normal form of the game G^*, but an even better course would be to construct a theory explicitly designed for extensive-form games.

Equilibrium Evolutionarily Stable Sets

An alternative approach to the nonexistence of an evolutionarily stable strategy is to generalize the ESS concept to a set-valued equilibrium concept. This approach has precedents in the refinements literature, where set-valued equilibrium concepts have been proposed as refinements of Nash equilibrium. The technique in both the refinements and evolutionary literature, when confronted with a case in which no single strategy profile exhibits all of the properties we would like, is to expand the set of strategies under consideration until the set as a whole exhibits the desired properties.

Set-valued equilibrium notions may do a better job than single-valued notions of capturing the behavior in which we are interested. As we have seen, a strategy in an extensive-form game that leaves an information set unreached will not be an ESS. But if we do not care what action is taken at this information set, as long as it does not affect the path of play, then the interesting feature of behavior is characterized by a set-valued equilibrium concept that is not specific about what happens at the unreached information set.[17]

How are we to interpret a set-valued solution concept, and how much information does it provide? This clearly depends upon the characteristics of the set. For example, we can always simply take the entire state space to be our set-valued solution. Although this ensures that existence will never be a problem, it is also not particularly informative. The most convenient property for a set-valued solution concept is for the members of the set to correspond to a single outcome in the underlying game because the behavioral implications of the solution would then be unambiguous; we shall be especially interested in such cases.

We have reason to believe that set-valued equilibrium concepts may be particularly relevant in an evolutionary context. The selection or learning mechanism driving the evolutionary process may be effective at shaping behavior at information sets that are often reached, but may not be able to bring any pressure to bear on actions at unreached information sets. If this is the case, then we cannot hope for more than a set of outcomes that differ in the behavior prescribed at unreached

---

17. Refinements of the Nash equilibrium concept such as Kohlberg and Mertens's strategic stability [135] are in this vein. See van Damme [241] for a discussion of refinements that addresses this issue.

information sets, though in the most convenient case the forces shaping behavior produce the same behavior at each of the outcomes in this set.

The first set-valued equilibrium notion to be discussed here is the idea of an *equilibrium evolutionarily stable set*, offered by Swinkels [230]:

Definition 2.6   Let $C(\sigma) = (C(\sigma_I), C(\sigma_{II})) \subset S_I \times S_{II}$ be the support of $\sigma = (\sigma_I, \sigma_{II})$, and let $B(\sigma) \subset \Delta S \times \Delta S$ be the set of all strategies that are best replies to σ. Then a set of strategies $X \subset \Delta S \times \Delta S$ is *equilibrium evolutionarily stable* if X is closed, each $\sigma \in X$ is a Nash equilibrium, and there exists $\epsilon^* > 0$ such that $\forall \epsilon \in (0, \epsilon^*), \forall \sigma \in X, \forall \sigma' \in \Delta S \times \Delta S$,

$$C(\sigma') \subset B((1 - \epsilon)\sigma + \epsilon\sigma') \Rightarrow (1 - \epsilon)\sigma + \epsilon\sigma' \in X. \tag{2.9}$$

To interpret this condition, we think of σ' as being a mutant. Then $(1 - \epsilon)\sigma + \epsilon\sigma'$ is the state of the population after the mutant σ' has entered the population. The left side of (2.9) is the statement the mutant σ' is a best response to the *postentry* environment that will exist after the mutant has entered. Condition (2.9) requires that if we can create a population state by adding mutants to the existing population who will be best responses to the resulting state, then that state must be contained in the equilibrium evolutionarily stable set. We can thus intuitively think of an equilibrium evolutionarily stable set as containing every strategy profile to which the system can shift via a process of adding best-response mutants.

We can interpret equilibrium evolutionarily stable sets in two ways. First, even though the agents in an evolutionary model are often viewed as unthinking automata, we might think of mutants as calculating individuals who assess the current state of the game before choosing their actions. These mutants then realize that they should choose a strategy that is a best reply to the environment in which they will find themselves after they have chosen their strategies, and we need not worry about mutants who do not choose best replies. Alternatively, it may be that mutants are more vulnerable to evolutionary pressures than existing strategies. In a learning interpretation, this may be because those who have just entered the game or have just chosen new strategies are especially likely to assess the performance of their strategies and make appropriate adjustments. As a result, mutants who do not play best replies quickly change their behavior, before they can have an impact on the population, and we again need only worry about best-reply mutants.

In G^*, equilibrium evolutionarily stable sets have the attractive feature of corresponding to a single outcome:

Remark 2.4 Game G^* has a unique equilibrium evolutionarily stable set in which player II always plays R, yielding the subgame-perfect equilibrium outcome.

In particular, the strategy profile $(L\ell, L)$ cannot be contained in an equilibrium evolutionarily stable set because the set would also have to contain the strategy profile in which player I chooses $L\ell$ and player II mixes equally between L and R, at which point there are mutant strategies (placing higher probabilities on R for player II) that pass the test of being best replies to their postentry environment, but take the system outside the set of Nash equilibria.[18]

Nonequilibrium Evolutionarily Stable Sets

A different outcome is provided by Sobel's nonequilibrium evolutionarily stable sets [223].[19] The common assumption in evolutionary analyses is that the populations are large, so that we can treat them as being essentially infinite, in the sense that sampling with and without replacement gives equivalent results. Sobel, however, explicitly assumes that agents are drawn from two finite populations, one of player I's and one of player II's. We then have

Definition 2.7 Strategy profile σ, identifying a pure strategy for each agent of each population, can *replace* σ' if σ and σ' differ only in the strategy of a single agent, with that agent's expected payoff against a randomly selected agent of the opposing population being at least as high under σ as it is under σ'. A set X is a *nonequilibrium evolutionarily stable* set if it is nonempty and minimal with respect to the property that $\{\sigma' \in X$ and σ able to replace $\sigma'\}$ implies $\sigma \in X$.

---

18. Thomas [234, 235] offers the related set-valued equilibrium concept of an *evolutionarily stable* set, obtained by directly extending the definition of an ESS to sets. An evolutionarily stable set differs from an equilibrium evolutionarily stable set in that *any* strategy in a neighborhood of an evolutionarily stable set that does as least as well as the strategies in the set must also be in the set, whereas equilibrium evolutionarily stable sets impose a similar requirement only for strategies that are *best responses* in the neighborhood. The unique evolutionarily stable set in game G^* coincides with the equilibrium evolutionarily stable set. See Balkenborg [6], Balkenborg and Schlag [7, 8], and Schlag [204, 205] for further development.

19. Predecessors of Sobel include Kim and Sobel [133] and Blume, Kim, and Sobel [34]. Related ideas appear in the formulation of *cyclically stable sets* by Gilboa and Matsui [99] and Matsui [148].

Hence a nonequilibrium evolutionarily stable set must contain any population profile to which it can move via a process of having one player at a time shift in a direction that does not decrease the player's payoff. It follows immediately from the definition that

Remark 2.5   In Game G^*, there is a single nonequilibrium evolutionarily stable set, containing (among other strategy profiles) every strategy profile in which player I chooses any mixture between $L\ell$ and Lr as well as every strategy profile in which player I mixes between $R\ell$ and Rr, while player II chooses R.

Both of the Nash equilibrium outcomes of G^* are included in the nonequilibrium evolutionary state set, as are some profiles that are not equilibria. To see how this occurs, consider a state in which all agents play their part of the subgame-perfect equilibrium $(R\ell, R)$. This state can be replaced by a state in which a member of population I plays Rr because the shift from ℓ to r does not decrease the agent's payoff. This state can in turn be replaced by one in which yet another agent in population I plays Rr; continuing in this way, we can successively replace states until we have all agents in population I paying Rr. At this point, we have a state that does not correspond to a Nash equilibrium because agents in population II are playing R but L is a best reply. However, this state must be contained in the nonequilibrium evolutionarily stable set, at least if the latter is to contain the subgame-perfect equilibrium, because we have moved to the current state from the subgame-perfect equilibrium via a series of replacements. Now additional replacements can occur that cause agents in population II to switch from R to L because such switches increase the payoffs of these agents (given that agents in population I play Rr). These replacements again yield states that are not equilibria; this can continue until we reach a state in which agents in population I play Rr and those in population II play L. Further replacements can then cause agents in population I to switch to L, which is now a payoff-increasing switch, and can lead to a state in which agents in population I play $L\ell$, while those in population II play L. We have then reached a Nash equilibrium that is not subgame-perfect.

Dynamic Interpretations

The list of alternative notions of evolutionary stability has been steadily growing. What accounts for the differences in these various stability notions? I think the most revealing way to assess these notions is to recognize that behind each concept lies an implicitly dynamic model, just as many conventional equilibrium refinements are motivated by implicitly dynamic models of what players will do when confronted with an out-of-equilibrium event. These dynamic models differ most noticeably in the different provisions they make for the treatment of mutants.

To see the differing dynamic foundations lying behind static equilibrium concepts, compare the concepts of a weak evolutionarily stable strategy and equilibrium evolutionarily stable set, both of which support the subgame-perfect outcome. The weak evolutionarily stable strategy concept assumes that mutations switching agents from strategy $R\ell$ to Rr are immediately repulsed. Hence attention can be limited to the single strategy profile $(R\ell, R)$, which appears to involve an assumption to the effect that perturbations causing player II to play L occur frequently enough to eradicate any switches from $R\ell$ to Rr. In contrast, the concept of an equilibrium evolutionarily stable set allows mutations introducing Rr to proceed unchecked until the boundary is approached at which R ceases to be a best reply for player II. Here, further mutations are impossible. The assumption appears to be that it is not perturbations but best-reply behavior that disciplines mutations. As long as R is a best reply for player II, there is no discipline imposed on mutations toward Rr. Once the edge is reached at which L is a best reply, however, then a rush of player-II agents to L pushes player I back to $R\ell$, which in turn makes R once again a best reply for player II.

In light of this interpretation, consider the difference between equilibrium and nonequilibrium evolutionarily stable sets. Attention is again directed at the treatment of mutants in the dynamic models that implicitly lie behind the two concepts. Begin with the subgame-perfect equilibrium $(R\ell, R)$ in game G^*. Both the equilibrium and nonequilibrium evolutionarily stable set concepts allow mutations to switch agents in population I toward Rr, to the point that L can become a weak best reply for player II. Here, however, the two part company. As we have seen, one interpretation of an equilibrium evolutionarily stable set is that mutants adjust faster than incumbents. Any further drift toward Rr now prompts player II to adjust toward the best reply of L, which *immediately* pushes the mutant agent I's back to $R\ell$ (and then player II back to R) and restores the Nash equilibrium. A nonequilibrium evolutionarily stable set, in contrast, admits the possibility that once sufficient mutation toward Rr has occurred to make L a best reply for player II, the primary adjustments may first be for player-*II* agents to switch to *L*, with this movement continuing until player-*I* agents no longer find it a best reply to play *R* at their first move and hence switch to *Lℓ* or *Lr*, leading away from the subgame-perfect equilibrium.

Despite the contrast in the terms, namely, *equilibrium* evolutionarily stable sets and *non*equilibrium evolutionarily stable sets, these are quite similarly motivated and constructed concepts. Lying behind both is a presumption that agents in the population flow toward higher-payoff strategies and away from lower-payoff strategies, but can be switched by mutations among equal-payoff strategies. Starting from the subgame-perfect equilibrium, both agree that mutations from *Rℓ* to *Rr* can occur. They differ only on what happens when these mutations lead the system to the brink of no longer being a Nash equilibrium, where they make different assumptions as to the relative rates at which agents in the two populations adjust their strategies. But this seemingly subtle difference leads one concept to restrict attention to the subgame-perfect equilibrium, while leading the other to include other Nash equilibria and nonequilibrium outcomes as well.

2.3   Dynamic Systems

The finding that the various stability notions, offered as alternatives to the concept of an evolutionarily stable strategy, differ in their implicit, underlying dynamics, directs our attention to dynamic evolutionary models. I believe that the study of such models will play a central role in assessing evolutionary stability concepts.

Because biologists working with evolutionary models have a clear idea of how the evolutionary process operates, it is only natural that they would investigate directly the dynamic process whose outcome the ESS is supposed to describe. With the help of some simplifying assumptions, the resulting models lead to the *replicator dynamics*, a system of deterministic difference or differential equations that plays a central role in biological models.

Economists have similarly studied dynamic models, both in an effort to motivate solution concepts such as ESS, and its alternatives, and in response to the observation that an ESS often fails to exist. In many cases, economists have simply worked with the replicator dynamics. Given their original biological motivation, it is not obvious that the replicator dynamics should command attention in economic applications. However, several authors have recently noted that the replicator dynamics can emerge from simple learning models. Although the learning models in question are stylized and arbitrary in many respects, and certainly provide no reason to claim that the replicator dynamics are in any sense the "right" model, they do suggest that the replicator should not be automatically dismissed in economic applications.[20] The process by which agents choose their strategies in a learning context is more likely to resemble the replicator dynamics if imitation is an important component of the learning process; a process driven by imitation shares many of the features of biological reproduction, with agents who are successful giving rise to additional agents playing the same strategy.

This section presents the basic results concerning the replicator dynamics and their relationship to evolutionarily stable strategies.[21] Economists have also examined a wide variety of dynamic evolutionary models using other techniques. This section closes with a brief discussion of a frequently used model whose techniques of analysis and outcome provide a sharp contrast with those of models based on the replicator dynamics—the Markov model of Kandori, Mailath, and Rob [128].

Replicator Dynamics

In discussing the replicator dynamics, I shall often interpret the system in biological terms. Consider first the most common model, containing a single large population of agents who will be matched in pairs to play a symmetric game. Unlike the case of an evolutionarily stable strategy, each agent is now associated with a pure strategy. The counterpart of a mixed strategy in this model is a polymorphic population in which different agents play different pure strategies. One could easily allow the agents to play mixed strategies as long as only a finite number of such strategies could occur in the population at any single time, whereas allowing an infinite number of different

---

20. One such derivation is presented in chapter 3. See also Börgers and Sarin [47], Cabrales [54], and Schlag [206]. Cressman [69] shows that if a state is a hyperbolic stationary state under both the replicator dynamics and an arbitrary dynamic from a more general class of "monotonic" dynamics (that includes the replicator dynamics), then it is a source, saddle, or sink under the monotonic dynamic if and only if it is a source, saddle, or sink under the replicator dynamics. This is a reflection of the fact that the properties of stationary states are relatively insensitive to alterations in the specification of a dynamic system that preserve the qualitative nature of its local behavior. The implication is that we can often learn much from studying the replicator dynamics.
21. This material is well covered by Weibull [248] and by van Damme [240], as well as by Hofbauer and Sigmund [119], and the treatment here will be brief.

strategies requires more advanced techniques. Notice that a model in which agents play pure strategies and mixed-strategy outcomes appear as population polymorphisms is similar in spirit to Harsanyi's purification model of mixed strategies [115].

Let $N_i(t)$ be the number of agents playing pure strategy i at time t and let $N(t)$ be the total number of agents at time t. Let $x_i(t)$ be the proportion of agents playing pure strategy i at time t, so $x_i(t) = N_i(t)/N(t)$. Let $x(t)$ be the vector of proportions playing the various pure strategies. $x(t)$ is then a member of the mixed-strategy simplex, and we will often interpret $x(t)$ as a mixed strategy.

There are several versions of the replicator dynamics and even more methods for deriving and motivating these versions. First, consider discrete-time models. Van Damme [240, section 9.4] suggests the following argument, based on a nonoverlapping generations model of the population.[22] In each period, agents in the population are paired at random to each play exactly one game. At the end of the period, each agent is replaced by agents who play the same strategy and whose number is given by the payoff the original agent received in the game played in that period. Then let the expected number of replacements for each agent playing strategy i be denoted by $\pi(i, x)$ and let the expected number of agents next period be $\sum_{i \in S} x_i \pi(i, x)$. For convenience, let the average payoff in the population be denoted by $\bar{\pi}$, where

$$\sum_{i \in S} x_i \pi(i, x) = \bar{\pi}(x).$$

If we assume that actual outcomes will match expected outcomes, then the proportion of the population playing strategy i next period will be

$$x_i(t + 1) = x_i(t) \frac{\pi(i, x(t))}{\bar{\pi}(x(t))}.$$

We can than represent the evolution of the population as

$$x_i(t + 1) - x_i(t) = x_i(t) \frac{\pi(i, x(t)) - \bar{\pi}(x(t))}{\bar{\pi}(x(t))}. \tag{2.10}$$

This discrete replicator dynamic is examined by Dekel and Scotchmer [72], Hofbauer and Sigmund [119], Maynard Smith [149, p. 183], and van Damme [240]. The assumption that actual and expected outcomes

---

22. Hofbauer and Sigmund [119, p. 13] present a similar argument.

will coincide is commonly motivated by observing that the population is assumed to be large, so that one can appeal to the law of large numbers. The next chapter returns to this issue.

The derivation of (2.10) is based on the assumption that generations do not overlap, so that all agents reproduce at the same time and parents do not survive the reproduction process.[23] For some biological applications, where reproduction occurs in a fixed breeding season and organisms live only from birth to the next breeding season, this may be the most appropriate model. In other applications, in which reproduction is not synchronized, an overlapping generations model may be more appropriate. This may especially be the case in the learning applications that interest us, where there may be no reason for all agents to assess their strategies at the same time, and hence agents who have just selected a new strategy will coexist with others who have maintained their current strategy for some time.

Binmore [19] examines the overlapping generations case; a similar derivation appears in Taylor and Jonker [232]. Suppose that in each period of time of length τ, a fraction τ of the population gives birth, with an individual playing strategy i giving birth to $\pi(i, x)$ offspring who also play strategy i. Then the expected number of agents playing strategy i in period $t + \tau$ is given by $N_i(t + \tau) = N_i(t) + \tau N_i(t)\pi(i, x(t))$. Dividing each side by the total number of agents next period (given by $\sum_{j \in S} N_j(t)(1 + \tau\pi(j, x(t)))$ and then dividing the numerator and denominator of the right side by the total number of agents this period, we have

$$x_i(t + \tau) = \frac{x_i(t)(1 + \tau\pi(i, x(t)))}{\sum_{j \in S} x_j(t)(1 + \tau\pi(j, x(t)))}$$

and hence

$$x_i(t + \tau) - x_i(t) = x_i(t)\frac{\tau\pi(i, x(t)) - \tau\overline{\pi}(x(t))}{1 + \tau\overline{\pi}(x(t))}. \tag{2.11}$$

Equations (2.10) and (2.11) are the two most common discrete replicator dynamics. Cabrales and Sobel [55] consider a variation of (2.11) given by

$$x_i(t + \tau) - x_i(t) = x_i(t)\frac{\tau\pi(i, x(t)) - \tau\overline{\pi}(x(t))}{\overline{\pi}(x(t))}. \tag{2.12}$$

---

23. Mailath [145, p. 268] notes the importance of the nonoverlapping generations assumption.

Hofbauer and Sigmund [119, p. 133], considering the case of $\tau = 1$, suggest a version that includes (2.11) as a special case:

$$x_i(t + \tau) - x_i(t) = x_i(t) \frac{\pi(i, x(t)) - \overline{\pi}(x(t))}{C + \overline{\pi}(x(t))}. \tag{2.13}$$

The constant C is interpreted as the rate of "background fitness," which is the rate of fitness that would prevail in the absence of the strategic interaction in question. Though background fitness notions have played only a minor role in economic applications of evolutionary game theory, similar notions will make an appearance in chapter 9.

It is often convenient to work in continuous time. Van Damme [240] invokes an assumption that time periods are very short to pass from (2.10) to

$$\frac{dx_i}{dt} = x_i \frac{\pi(i, x) - \overline{\pi}(x)}{\overline{\pi}(x)}. \tag{2.14}$$

van Damme then observes that by a rescaling of time, this can be converted to the following system, which has identical solution trajectories:

$$\frac{dx_i}{dt} = x_i(\pi(i, x) - \overline{\pi}(x)). \tag{2.15}$$

Alternatively, one can obtain (2.14) directly by taking the limit as the length of a time period τ approaches zero in (2.12), and can obtain (2.15) directly by taking the limit $\tau \to 0$ in (2.11) (or (2.13) with $C = 1$).[24]

In many cases, it does not matter which replicator dynamic we use. However, Hofbauer and Sigmund [119, p. 133] note that with $\tau = 1$, the outcome of (2.13) can depend on C. This in turn indicates that (2.10) and (2.11) can give different results. The results in Samuelson and Zhang [202] and in Dekel and Scotchmer [72] show that (2.10) and (2.15) can give different outcomes. Cabrales and Sobel [55] and Weibull [248] explore the forces behind this difference. Boylan [49] shows that (2.10) and either of (2.14) or (2.15) can give different results. Finally, Bomze and Eichberger [41] and Hofbauer and Sigmund [119, p. 273] note that the transformation leading from (2.14) to (2.15) can break down when there is more than one population, and that the two continuous dynamic systems can give different outcomes

---

24. Hofbauer and Sigmund [119, section 16.1] and Weibull [248, chapter 3] provide direct derivations of (2.15) without the intermediate step of a discrete-time model.

in this case. Some of these differences are explored further by Seymour [215].[25]

Equilibrium

Once a formulation for the replicator dynamics is chosen, what can we learn from it? The replicator dynamics are related to the ESS concept as well as to the Nash equilibrium concept. In exploring these relations, I restrict attention to the case that has dominated the literature, especially the biological literature, namely, symmetric games played by a single population of players and with the replicator dynamic given by (2.15).

Let $x(x_0, t): \Delta S \times \mathbb{R} \to \Delta S$ denote a solution to (2.15). The interpretation is that $x(x_0, t)$ gives the state of the population at time t given that its state at time zero, or initial condition, is x_0. The right side of (2.15) is a polynomial in the population shares, and hence is Lipschitz-continuous. This in turn suffices to ensure that for any initial condition x_0, there is a unique solution $x(x_0, t)$.[26] Given a state x_0, the set $\{x(x_0, t) : t \in \mathbb{R}\}$ is the *trajectory* or the *orbit* passing through x_0. This trajectory is *periodic* with period t if there exists a time t with $x(x_0, t) = x_0$. A set X is invariant if $x_0 \in X$ implies that $x(x_0, t) \in X$ for all $t \in \mathbb{R}$. The set X is forward-invariant if $x_0 \in X$ implies that $x(x_0, t) \in X$ for all $t \in \mathbb{R}_+$.

Because the selection mechanism lying behind the replicator dynamics is based on reproduction, strategies that are currently not played by any agent in the population cannot be introduced into the population by the replicator dynamics. At the same time, it is evident from (2.15) that the growth rate of the proportion of the population playing any given strategy is finite. The proportion of the population playing a strategy can get very small, but never reaches zero. Literally speaking, there is no extinction, though cases can arise in which the proportion of the population playing a particular strategy approaches zero. As a result, the strategy simplex ΔS is invariant under the replicator dynamics. In addition, any boundary face (i.e., any set obtained by setting the population proportions of one or more strategies equal

---

25. Things become more complicated if one adds perturbations to the replicator dynamics. Foster and Young [79], Fudenberg and Harris [85], and Cabrales [54] examine models based on stochastic differential equations.

26. This is the Picard-Lindelöf theorem. See Hirsch and Smale [118] or Hale [111].

to zero) is also invariant. In the extreme, any monomorphic population state obtained by attaching all agents to the same strategy is
invariant.

We now require some stability notions. The following are standard:

Definition 2.8 The state $x^* \in X$ is a *stationary state* of the differential
equation $dx/dt = f(x)$ defined on $X \subset \mathbb{R}^n$ if $f(x^*) = 0$. x^* is *stable* if
it is a stationary point with the property that for every neighborhood
V of x^*, there exists a neighborhood $U \subset V$ with the property that if
$x_0 \in U$, then $x(x_0, t) \in V$ for all $t > 0$. And x^* is *asymptotically stable* if
it is stable and there exists a neighborhood W of x^* such that $x_0 \in W$
implies $\lim_{t\to\infty}(x_0, t) = x^*$.

Stationary states, also sometimes called "rest points" or "dynamic
equilibria," have the property that if the dynamic system begins at such
a state, then it remains there. However, a stationary state can still be unstable. For example, x^* may be stationary but may be a source, meaning
that for every initial condition close to x^*, the dynamic system moves
away from x^*. A stable state is a stationary point with the property
that if a trajectory starts nearby, then it stays nearby, though it need not
converge to x^*. An asymptotically stable state satisfies the additional
property that trajectories beginning from all nearby states converge
to x^*.[27]

A first familiar result is then

Proposition 2.5.1 If (x, x) is a Nash equilibrium of a symmetric game
G, then x is a stationary state of the replicator dynamic (2.15).

Proposition 2.5.2 If x is a stationary state of (2.15), (x, x) need not be
a Nash equilibrium of G.

Proposition 2.5.1 is easily established. A Nash equilibrium has the
property that all strategies played with positive probability receive the
same payoff, which must be as high as the payoff to any strategy that
does not receive positive probability. The right side of (2.15) must then
be zero for every strategy, either because the strategy is played with
zero probability ($x_i = 0$) or the strategy receives the same payoff as

---

27. Weibull [248] contains an example showing that the requirement that a stationary
state x^* be stable is not redundant in the definition of asymptotic stability. In particular,
one can find a stationary state x^* with the property that the trajectories from all nearby
initial conditions converge to x^*, but stray far away in the process, so that trajectories that
start nearby do not remain nearby.

every other, and hence receives the average payoff (giving $\pi(x_i, x) - \bar{\pi}(x) = 0$).

A state x can be stationary and yet (x, x) can fail to be a Nash equilibrium because the superior best replies that prevent (x, x) from being a Nash equilibrium are played by zero proportion of the population in state x. As an extreme example, recall that every monomorphic population state is a stationary state of the replicator dynamics, even though some of these states may not be Nash equilibria.

The ability of a point x to be stationary without (x, x) being a Nash equilibrium, because all of the strategies that are better replies are absent from the population and hence cannot increase in proportion, exploits one of the less attractive features of the replicator dynamics, namely their inability to introduce new strategies into the population. Imitation may be important, but if payoffs can be increased by introducing a currently unused strategy into play, then we would expect some agent to eventually choose such a strategy, either by design or by mistake. As a result, it is important to investigate how outcomes of the replicator dynamics fare when currently unused strategies are injected into the population. A first step is to examine stability.

A natural conjecture is that a stationary state that is not a Nash equilibrium cannot be stable. Instead, any perturbation that introduces a better reply than the existing strategy into the population, no matter how small, must cause the better reply to increase in proportion and hence must lead away from the original state. Making this precise gives

Proposition 2.6 If x is stable, then (x, x) is a Nash equilibrium; but there exist Nash equilibria (x, x) such that x is not stable.

Theorems establishing some form of the claim that "stability implies Nash" appear throughout evolutionary game theory. In conventional game theory, the conditions for a Nash equilibrium are often described as being necessary for behavior to be self-enforcing because otherwise some agents would have incentives to alter their behavior. Proposition 2.6 and related results provide a version of this necessity in a dynamic model.

The argument establishing that stable states are Nash equilibria is straightforward, exploiting the fact that if better replies exist, then perturbations introducing those best replies must yield dynamics that lead away, though the proof requires some attention to detail. To show

	s_1	s_2
s_1	1, 1	0, 0
s_2	0, 0	0, 0

Figure 2.5
Unstable Nash equilibrium, (s_2, s_2)

that a Nash equilibrium need not be stable, van Damme [240, corollary 9.4.5] notes that we need only examine a Nash equilibrium in dominated strategies, as in figure 2.5. The strategy profile (s_2, s_2) is a Nash equilibrium in this game, but any trajectory starting from an initial condition in which the proportion of the population playing strategy s_1 is nonzero must converge to the state in which all agents play s_1.

The example of figure 2.5 may leave the impression that if we simply restricted attention to Nash equilibria in undominated strategies or to trembling-hand-perfect equilibria, then we would achieve stability. However, van Damme [240, p. 229] shows that one can add a third strategy to the game in figure 2.5 that causes (s_2, s_2) to be a perfect equilibrium.[28] Once this strategy is added, it is still the case that any initial condition where some agents play s_1 and none play the third strategy must converge to a state where all agents play s_1, precluding the stability of s_2.

A stable state also need not be a perfect equilibrium. For example, consider figure 2.3. Every state in which only strategies s_1 and s_2 are played is stable, but the only perfect equilibrium is one in which all agents play s_1. We can thus see that the connection between stability and refinements such as trembling hand perfection is not particularly strong:

Proposition 2.7 A stable state need not be a perfect equilibrium, nor need a perfect equilibrium be stable.

Figure 2.3 provides some interesting insight into the nature of stability. Consider a state, denoted by x, in which all agents play either strategy s_1 or s_2. Now consider a nearby state in which an equal proportion of both types of agents are switched to strategy s_3. The resulting

---

28. To do this, it suffices that the third strategy be a strictly inferior reply to s_2 and that s_2 fare better than s_1 against the third strategy.

replicator dynamics trajectory will (asymptotically) eliminate the strategy s_3, decrease the proportion of agents playing s_2, and increase the proportion playing s_1. These dynamics do not return the system to the original state x, but such a return is not required in order for us to pronounce x stable. Instead, stability requires only that as long as the initial perturbation away from x is not too large the system not stray too far from x, and this condition is satisfied in this case.

A state in which a population playing the game in figure 2.3 is split between strategies s_1 and s_2 may be stable, but it is not asymptotically stable. One might conjecture that if stability is increased to asymptotic stability, then we would recover perfection. We have

Proposition 2.8 If x is asymptotically stable, then (x, x) is a perfect and isolated equilibrium. However, neither perfect nor perfect and isolated equilibria need be asymptotically stable.

Bomze [39] proves that asymptotically stable states are perfect and isolated. Figure 2.3 provides a perfect equilibrium, in which all agents play s_1, that is not asymptotically stable. Asymptotic stability fails in this case because there are arbitrarily close states where almost all agents play s_1 but some play s_2 that are also stationary states of the dynamics, and hence from which the dynamics do not converge to s_1. However, such states exist because the perfect equilibrium in this case is not isolated. To see that an isolated, perfect equilibrium need not be asymptotically stable, we need only recall that the game in figure 2.5 can be augmented by a third strategy to produce an isolated, perfect equilibrium, namely (s_2, s_2), that is not even stable.

The finding that asymptotically stable equilibria are isolated and perfect recalls the similar characterization of evolutionarily stable strategies. What is the relationship between the two? Taylor and Jonker [232] and Zeeman [253] show the following:[29]

Proposition 2.9.1 Every ESS is asymptotically stable, but asymptotically stable states need not be ESSs.

---

29. Hopkins [121] shows that for the class of "positive definite" dynamic processes (which includes the replicator) on symmetric games, any regular ESS is asymptotically stable, thus extending proposition 2.9.1 for regular ESSs. (Hopkins defines an ESS to be regular if it contains all best replies within its support, a property also sometimes called "quasi-strictness.") Bomze and Weibull [44] establish a counterpart of proposition 2.9.1 for neutrally stable stategies and stable states of the replicator dynamics.

	s_1	s_2	s_3
s_1	0, 0	1, −2	1, 1
s_2	−2, 1	0, 0	4, 1
s_3	1, 1	1, 4	0, 0

Figure 2.6
Asymptotically stable state that is not an ESS

Proposition 2.9.2   In two-player 2×2 games, a strategy x is an ESS if and only if the state x is asymptotically stable.

An example of a state that is asymptotically stable but is not an ESS is given by van Damme [240, p. 232]. The corresponding game is shown in figure 2.6. This game has a unique symmetric Nash equilibrium, given by $\sigma^* = (\frac{1}{3}, \frac{1}{3}, \frac{1}{3})$, that is perfect and isolated. Van Damme shows that this equilibrium is asymptotically stable. To see that it is not an ESS, notice that the equilibrium payoff from σ^* is $\frac{2}{3}$. If we let $\sigma = (0, \frac{1}{2}, \frac{1}{2})$, then we have $\pi(\sigma, \sigma^*) = \frac{2}{3} = \pi(\sigma^*, \sigma^*)$ but $\pi(\sigma, \sigma) = \frac{5}{4} > \pi(\sigma^*, \sigma) = \frac{7}{6}$. Hence σ can invade and σ^* is not an ESS.

To be evolutionarily stable is thus sufficient but not necessary to be asymptotically stable. Our initial thought might be that this relationship should be reversed, with asymptotic stability being the more demanding concept, because evolutionary stability requires only that the state in question be capable of repelling mutant invaders who appear one at a time, whereas asymptotic stability requires the state to exhibit stability properties in the face of all manner of (nearby) population polymorphisms. However, this seeming weakness of evolutionary stability is overwhelmed by the fact that the mutants an ESS is required to repel can play any mixed strategy, and once a mixed-strategy mutant has appeared, the system is given only the option of reducing the proportion of the population playing the mutant while being precluded from making any other adjustments.

To see the implications of this restriction, suppose that a population of agents playing $(\frac{1}{3}, \frac{1}{3}, \frac{1}{3})$ in figure 2.6 is invaded by a mutant bridgehead of $(0, \frac{1}{2}, \frac{1}{2})$. When searching for an evolutionarily stable strategy, we interpret this as a monomorphic group of agents playing the mixed strategy $\sigma^* = (\frac{1}{3}, \frac{1}{3}, \frac{1}{3})$ and another, small group playing the mixture $\sigma = (0, \frac{1}{2}, \frac{1}{2})$. The only adjustment that is now allowed is to alter the proportions of the population playing these two mixed strategies. We

	σ^*	σ
σ^*	$\frac{2}{3}, \frac{2}{3}$	$\frac{7}{6}, \frac{2}{3}$
σ	$\frac{2}{3}, \frac{7}{6}$	$\frac{5}{4}, \frac{5}{4}$

Figure 2.7
Derived game between candidate ESS and mutant

then essentially have a two-strategy game, involving σ^* and σ, with payoffs given in figure 2.7. The strategy σ is dominant in this game, causing σ to push σ^* to extinction and hence ensuring that the latter is not an ESS.

This can be contrasted with the case of the replicator dynamics. Here an initial state characterized by σ^* is viewed as a polymorphic population where one-third of the population is playing each of the pure strategies s_1, s_2, and s_3. The introduction of the mutant σ is interpreted as a perturbation that increases the proportions of the population playing strategies s_2 and s_3 by like amounts, while reducing the proportion playing s_1. The replicator dynamics are now free to make whatever adjustments in the population proportions are induced by the payoffs attached to this new state. This perturbation decreases the payoff to s_3, increases the payoff to s_1, and increases the payoff to s_2 even more, causing the replicator dynamics to increase the proportion of the population playing s_2 at the expense of s_3. As this movement continues, however, the payoffs of the three strategies change. The relative payoff to strategy s_1 will increase to such an extent that the proportion of the population playing s_1 will increase. This in turn can increase the relative payoff to s_3 to the extent that the proportion playing strategy s_3 begins to increase. This process continues until the replicator dynamics spiral back into state σ^*. Such a spiral is implicitly excluded by the ESS calculation.

This difference in the behavior of the replicator dynamics and the ESS concept depends crucially on the existence of at least three strategies. Suppose that a mixture σ^* is asymptotically stable under the replicator dynamics in a game with two pure strategies, s_1 and s_2. A perturbation that leads to a state σ where the proportion playing strategy s_2 is increased must lead to a relatively higher payoff for strategy s_1, in order for the replicator dynamics to push the system back toward σ^*. But this increased payoff for strategy s_1 ensures that σ^* is a better reply than is

σ (because σ^* places more weight on s_1), which is the requirement for σ^* to be evolutionarily stable. This suggests that evolutionary stability and asymptotic stability in the replicator dynamics are equivalent games with only two strategies, as is confirmed by proposition 2.9.2.

The distinction between asymptotic stability and evolutionary stability thus arises out of the richer dynamic forces that can appear when there are three or more strategies. We can see this in the results of Bomze and van Damme [43], who provide a characterization of evolutionarily stable strategies in terms of dynamic systems. In particular, fix a symmetric game G with a finite number of strategies and fix a strategy σ^*, where σ^* may be a pure or mixed strategy. For any alternative pure or mixed strategy $\sigma \neq \sigma^*$, we consider a "derived" 2×2 game $G_{(\sigma^*, \sigma)}$. This is a game in which there are two strategies, σ^* and σ, and whose payoff function is given by restricting the payoff function of game G to these two strategies. An example of such a derived game is given in figure 2.7. Bomze and van Damme [43, theorem 1] show

Proposition 2.10   State σ^* is evolutionarily stable in a symmetric game G if and only if for every strategy $\sigma \neq \sigma^*$, the state in which all agents play σ^* is asymptotically stable under the replicator dynamic (2.15) in the derived game $G_{(\sigma^*, \sigma)}$.

These results are encouraging. Asymptotically stable states have some equilibrium refinement properties. When an ESS exists, then it corresponds to an asymptotically stable state, providing a convenient algorithm for finding the latter. In addition, asymptotically stable states can exist even when an ESS does not, potentially allowing some relief from the existence problems of evolutionarily stable strategies.

How much relief has been obtained from the possibility that an ESS might not exist? In answering this question, it is helpful to move beyond the symmetric games, with which this section has been concerned thus far, to asymmetric games. The existence problems with evolutionary stable strategies become especially apparent in asymmetric games. How do the replicator dynamics fare in such games? Strong results for this case have been established by Ritzberger and Vogelsberger [189] and by Ritzberger and Weibull [190].[30] In particular, let σ now be a population profile for an asymmetric game, so that σ identifies, for each

---

30. Ritzberger [188] obtains similar results. Hofbauer and Sigmund [119, p. 282, exercise 1] note that asymptotically stable states cannot be completely mixed.

population, the proportion of that population playing each of the available pure strategies. Then, according to Ritzberger and Vogelsberger [189, proposition 1] and to Ritzberger and Weibull [190, theorem 1]:

Proposition 2.11   A stationary point σ of an asymmetric game G is asymptotically stable under the replicator dynamic (2.15) if and only if σ is a strict Nash equilibrium of G.

An investigation of the replicator dynamics thus leads to results reminiscent of those that emerge from consideration of evolutionarily stable strategies. Asymptotically stable states are Nash equilibria that possess refinement properties. However, asymptotically stable states often fail to exist. The possibility of nonexistence is especially apparent in asymmetric games, where only strict Nash equilibria can be asymptotically stable. If we are to seek relief from the nonexistence of asymptotically stable states or the requirement of a strict Nash equilibria, we must follow the lead of those who have sought alternatives to the evolutionarily stable strategy concept, examining either alternative stability notions or alternative dynamic models.

Microfoundations

It will be helpful to briefly consider the possibility of alternative dynamic models.[31] The biological analogies that lie behind the replicator dynamics are often helpful, but we are especially interested in how people learn to play games. Suppose we begin with a model of how the agents in a game make and revise their choices. Will the resulting dynamic system bear any relationship to the replicator dynamics or similar systems of equations? What will be the equilibrium implications?

The work of Boylan [49, 51] is quite instructive in this regard, and also serves as a warning that choice models may give behavior that differs markedly from systems of difference or differential equations that seem to be derived from the models in a straightforward way. Consider the game in figure 2.8. Boylan examines a model in which this game is played by a finite population of agents. Each agent is matched to play exactly one game with a single opponent in each period, with

---

31. Swinkels [231], Ritzberger and Weibull [190], Balkenborg [6], Blume [33], and Hurkens [123] examine alternative stability notions, examining sets of outcomes with the property that the set as a whole satisfies a version of asymptotic stability.

	s_1	s_2	s_3
s_1	1, 1	2, 0	0, 2
s_2	0, 2	1, 1	2, 0
s_3	2, 0	0, 2	1, 1

Figure 2.8
Boylan's Game

each of the finite number of possible matching configurations being equally likely. After each such match, the agents in that match die and are replaced by two new agents. The payoffs of the game identify the strategies played by these new agents. For example, a match between an s_2 and an s_3 agent gives the payoffs $(2, 0)$, meaning that two of the new agents play strategy s_2 (and none plays s_3). A more economic interpretation of this process would view the agents not as dying and being replaced by new agents but rather as either retaining their own strategies or switching to the strategy of their opponent in their most recent match. Notice that the population size remains constant. We thus have a finite stochastic process.

Boylan [49] derives a deterministic, discrete replicator dynamic from the limit of this model as the population size goes to infinity. Boylan [49, 51] also derives a continuous replicator dynamic from the limit as the population size goes to infinity and the length of a time period goes to zero. In each case, Boylan shows that these dynamic systems provide good approximations of the behavior of the stochastic process over any finite time interval.[32] At the same time, the limiting behaviors of these models are quite different. For any fixed population size, the limiting distribution of the underlying stochastic choice process attaches probability to at most three states, consisting of those states in which all population members play the same strategy. Hence the stochastic process always produces a monomorphic population state in which two strategies are extinct. The limiting outcomes of the discrete replicator dynamic attach positive probability to these three states but also to an infinite number of other states in which two strategies sur-

---

32. More precisely, fix an initial condition, a time T, and an $\epsilon > 0$. Then for population sizes sufficiently large, the realization of the stochastic process and the solution of the deterministic replicator equation at any time $t \in [0, T]$ are within ϵ of each other with probability at least $1 - \epsilon$.

vive and one is extinct. The continuous replicator dynamic produces cyclical behavior, with periodic orbits that pass through the initial condition. If all strategies are initially present, then no strategy becomes extinct.

The important point here is that examining what appears to be the obvious deterministic difference or differential equation can give outcomes quite different than the underlying stochastic process. The derivations of the various versions of the replicator dynamic discussed above involved a number of approximations, "smoothing" arguments, and limiting operations. It is clear from Boylan's results that these procedures are not innocuous. When working with a dynamic system, it is accordingly important to understand clearly the model from which the dynamic behavior emerges.

Boylan's analysis takes as its point of departure a model that identifies, for each pair of strategies played by the agents in a match, the number of offspring of each agent. One readily interprets this in biological terms, but economic applications are more strained. For the latter, we need a model of the learning process that converts experience in the game into decisions to adopt certain strategies. A number of such models have recently appeared. Most prominent among these are the models of Young [250] and of Kandori, Mailath, and Rob [128]. Their popularity has been fueled by the fact that they typically yield strong results with a minimum of analytical complication. The forces driving the models are similar, though they differ considerably in the details of their construction. It will be useful to briefly examine the simpler model of Kandori, Mailath, and Rob [128], hereafter called the "KMR" model; we shall further restrict our attention to the simplest form of this model.

The model begins with a symmetric normal-form game, played by a single population of N players, each of whom is characterized by a pure strategy. In each period, these players are matched in pairs to play the game an infinite number of times. At each iteration of play, each player is equally likely to be matched with each of the $N - 1$ opponents. As a result, there is no randomness in the average payoffs received in each period, with each player receiving the expected payoff of playing against the mixed strategy given by the proportions of opponents playing the various pure strategies.

At the end of each period, all players "learn," meaning that they change their strategy to the best response to the current population state.[33] It is not essential that this learning process be so rigid as to switch every agent to a best reply. The results would be unchanged if some agents are randomly drawn in each period to switch to a best reply. Similarly, the model could be built around a number of alternative specifications, as long as they increase the proportion of agents playing a best response.

After learning, "mutations" occur. Each player takes an independent draw from a random variable that produces the outcome "mutant" with probability λ. If the agent is not a mutant (probability $1 - \lambda$), then the agent retains her current strategy. If a mutant, then the agent randomly chooses a strategy, with each of the strategies being equally likely.

The KMR model yields a Markov process. In the most commonly studied application, the game has only two strategies. Let these be denoted by X and Y. The state space of the KMR model is the set $\{0, 1, \ldots, N\}$, where a state identifies the number of agents playing strategy X. The transition matrix for this process is strictly positive, meaning that the probability of moving from any given state to any other is positive. To verify this, one need only note that beginning in any state, there is a positive probability that after learning has occurred, all agents are drawn to be mutants and happen to make the strategy choices that yield the new state in question.

Because its transition matrix is strictly positive, the KMR model has a unique stationary distribution that is independent of initial conditions and that will attach positive probability to all states in the state space. However, Kandori, Mailath, and Rob [128] show that if the mutation probability λ is sufficiently small, then the stationary distribution will concentrate almost all of its probability on only a few states. They accordingly direct our attention to the limit of the stationary distribution as λ get small, which I refer to as the "limiting distribution." Young [250] and Kandori, Mailath, and Rob [128] show that this limit is often easily computed and often produces powerful results.

---

33. If there is more than one best reply and agents are already playing a best reply, then they do not change strategies. If they are not choosing a best reply, then a rule for selecting among the best replies must be imposed. The specifics of this rule typically do not have a significant impact on the results. The most common rule is to assume that agents randomly choose a best reply, with each possible best reply being equally likely. This discussion will consider only games with two pure strategies, so that the issue of choosing among multiple best replies does not arise.

	X	Y
X	5, 5	0, 4
Y	4, 0	2, 2

Figure 2.9
Stag-Hunt Game

How do the KMR model and a model based on deterministic differential equations, such as the replicator dynamics, compare? Consider the Stag-Hunt Game of figure 2.9. This is a symmetric game played by a single population of players, each of whom may choose either X or Y. The game has two Nash equilibria in pure strategies, (X, X) and (Y, Y). It also has a mixed Nash equilibrium in which both players use X with probability $\frac{2}{3}$. Harsanyi and Selten [116] call (X, X) the "payoff-dominant equilibrium" because it is Pareto-superior to the other Nash equilibria of the game, and they call (Y, Y) the "risk-dominant equilibrium." In a 2×2 symmetric game with two symmetric, strict Nash equilibria, the risk-dominant equilibrium is characterized by having the larger basin of attraction under either the best-reply or the replicator dynamics.

Which of the Nash equilibria in the Stag-Hunt Game should be selected? The mixed-strategy equilibrium is commonly dismissed, but the choice between the two pure-strategy equilibria has provoked much debate. Conventional equilibrium refinements typically shy away from choosing between strict Nash equilibria, and most of the debate has accordingly been conducted in an evolutionary context.[34]

The phase diagram for the replicator dynamics is shown in figure 2.10. The equilibria (X, X) and (Y, Y), or more precisely, the state in which all agents play X and the state in which all agents play Y, are both asymptotically stable with respect to the replicator dynamics. The same is true of any dynamic in which population shares are positively related to average payoffs, including best-reply dynamics. The outcome of the replicator dynamics then depends critically on their

---

34. Among nonevolutionary theories that do discriminate between the two strict Nash equilibria of this game, Harsanyi and Selten's equilibrium selection theory [116] proposes the payoff-dominant equilibrium as the rational choice in the Stag-Hunt Game. Anderlini's theory of Turing machines engaged in cheap talk [1] also selects the payoff-dominant equilibrium. Carlsson and van Damme [60, 61] show that the risk-dominant equilibrium is the only one that is robust to a specific type of uncertainty about payoffs.

All Y   All X

Figure 2.10
Phase diagram for Stag-Hunt Game

initial condition. If this initial condition lies in the basin of attraction of the equilibrium (X, X), then the dynamics will converge to (X, X). An initial condition in the basin of attraction of (Y, Y) yields convergence to (Y, Y).

In contrast to the replicator dynamics, the KMR model gives a strikingly single-minded result for the Stag-Hunt Game. In the limit as the mutation level becomes negligible, the stationary distribution of the KMR model places all its probability on the risk-dominant equilibrium (Y, Y). To see how this result arises, notice that for most population states, the result of the best-reply dynamics is to immediately push the system to one of its strict Nash equilibria.[35] If we examine the system just after learning has occurred, then all we observe are states in which all agents play X or all agents play Y because either X or Y was a best reply to the previous state and all agents have switched to that best reply. Suppose that all agents play X. Mutations may then switch some agents to Y, but as long as not too many switch to Y at one time, the learning process immediately leads all agents back to X. Occasionally, however, mutations switch enough agents to Y as to make Y a best reply. The learning process then leads all agents to play Y. The system now remains in this state, with occasional mutations switching some agents to X, and learning then leading all agents back to Y, until enough agents simultaneously mutate to X to make X a best reply and lead the system back to all X.

---

35. The exception that motivates the "most" in "most population states" arises when the population proportions of Y and X are so close to the probabilities of the mixed-strategy equilibrium of the game that all X agents will have Y as a best reply (because they do not play themselves, and hence face disproportionately many Y agents) and hence will switch to Y; while all Y agents similarly switch to X. If these switches have been made, however, and if the population is sufficiently large, then there will be more Y agents than X agents in the population because (Y, Y) is the risk-dominant equilibrium, and hence there are more X than Y agents in the mixed-strategy equilibrium, with the former having now switched to Y. This makes Y best reply for all agents, leading to a state in which Y is played by all. At worst, it can then take two steps for learning to reach a strict Nash equilibrium.

These isolated bursts of rare mutations will continually switch the system between the two ends of the state space. Where will the system spend most of its time, in the state where all agents play X or the state where all agents play Y? This depends upon the relative probabilities of switching between these two states. Because (Y, Y) is the risk-dominant equilibrium and hence has the larger basin of attraction, the number of mutations from Y to X required to switch the system from all-Y to all-X is more than the number of mutations from X to Y required to make the reverse switch. The system then spends more of its time in the state where all agents play Y than in the state where all play X. As the probability of a mutation gets small, the probability of switching from all-Y to all-X gets arbitrarily small relative to the probability of a switch in the reverse direction, and the system spends virtually all of its time in equilibrium, where all agents play Y.

How do we interpret these results? The search for the evolutionary foundations of game-theoretic solution concepts leads from the notion of an evolutionarily stable strategy to alternative notions of evolutionary stability to dynamic models of evolutionary processes. I believe dynamic models represent our best hope for making progress on understanding equilibrium concepts, although they also present some difficult puzzles. The commonly used technique of modeling the evolutionary process as a system of deterministic difference or differential equations may tell us little about equilibrium concepts other than that "strict Nash equilibria are good." We can attempt to probe deeper into these issues by modeling the choices made by the agents. However, if we do so by investigating the commonly used models of Young [250] and of Kandori, Mailath, and Rob [128], we find quite different results, including the finding that "not all strict Nash equilibria are good." What is the relationship between these two types of models, one based on difference or differential equations and one on perturbed best-response dynamics? Under what circumstances are they applicable? The exploration of these questions begins in the next chapter.

3   A Model of Evolution

This chapter explores the relationship between (1) an explicit selection model; (2) deterministic differential equations, such as the classical replicator dynamics; and (3) the mutation-counting methodology of Freidlin and Wentzell [81], on which the models of Kandori, Mailath, and Rob [128] and of Young [250] are based. The learning model involved in this exploration is deliberately kept as simple as possible in order to focus attention on the main points. More general models are considered in subsequent chapters.[1]

This chapter concludes that whether differential equations or mutation-counting methods are useful in studying the evolutionary processes depends primarily on the length of time over which one chooses to study the selection model. If the population is sufficiently large, then the replicator dynamics provide a good approximation of the behavior of the population along any finite interval of time. However, the replicator dynamics do not provide a good description of the model's limiting behavior, which is captured by the mutation-counting methods of Freidlin and Wentzell [81].

The replicator dynamics are often dismissed by economists as being of purely biological interest, while other models based on differential equations are sometimes criticized for not having been derived from models of individual behavior. The emergence of the replicator dynamics from the "Aspiration and Imitation" model considered in this chapter suggests that such dynamics should not be automatically

---

1. The Aspiration and Imitation model considered in this chapter is taken from Binmore, Gale, and Samuelson [21] and Binmore and Samuelson [24]. The consideration of lengths of time is taken from discussions that first appeared in Binmore and Samuelson [26] and were pursued in Binmore, Samuelson, and Vaughan [28], from which the remainder of the chapter is taken.

rejected as models of how people behave. At the same time, the Aspiration and Imitation model is quite special. Other models will give rise to other approximations and we have no reason to believe the replicator dynamics have any particular significance. The important point is that systems of deterministic differential equations, like the replicator dynamics, can serve as useful approximations for the behavior of evolutionary processes over long but finite intervals of time.

The model examined in this chapter is a symmetric one in which agents are drawn from a single population to play the game without knowing which role in the game they are filling. Chapter 2 showed that symmetric and asymmetric models can behave differently. For example, the difficulties with conventional approaches, such as static notions of evolutionary stability, are especially apparent in asymmetric games. In the chapters that follow, I shall often shift between symmetric and asymmetric models, choosing whichever most effectively makes the point, noting in the process any results that are tied to a particular case.

3.1   The Aspiration and Imitation Model

The learning model to be studied is based on a process of comparing realized payoffs to an aspiration payoff level and, as a result, sometimes choosing new strategies by imitating others.[2] Consider the symmetric 2×2 game $\hat{G}$ of figure 3.1. Let there be a single population containing N agents, where N is finite. Time is divided into discrete intervals of length τ. In each time period, an agent is characterized by the pure strategy X or Y that she is programmed to use in that period. In each period of length τ, pairs of agents are randomly drawn (independently

---

2. Binmore, Samuelson, and Vaughan [28] present a model that leads to similar behavior in a biological setting. The specification of how agents learn is both an important area of research and one that has recently attracted tremendous attention. A careful discussion of this literature would require a book of its own (such as Fudenberg and Levine [95]). I shall be content here to note that models of learning in games are examined by, among many others, Banerjee and Fudenberg [9], Bendor, Mookherjee, and Ray [12], Binmore, Gale, and Samuelson [21], Binmore and Samuelson [24], Blume [33], Bomze and Eichberger [41], Börgers and Sarin [47], Canning [56, 57, 58], Cheung and Friedman [62], Crawford [66], Eichberger, Haller, and Milne [73], Ellison [75], Foster and Vohra [77, 78], Fudenberg and Kreps [86, 87, 88, 89], Fudenberg and Levine [91, 92, 93, 94], Jordan [124, 125], Kalai and Lehrer [127], Kaniovski and Young [129], Milgrom and Roberts [155, 156], Nachbar [163], Roth and Erev [195, 196], Selten [213], and Stahl [227, 228].

	X	Y
X	A, A	C, B
Y	B, C	D, D

Figure 3.1
Game $\hat{G}$

and without replacement) to play the game, with each agent playing one game and with the various ways agents can be matched being equally probable.[3]

The payoffs A, B, C, and D that appear in the specification of game $\hat{G}$ given by figure 3.1 are taken to be *expected* payoffs. *Realized* payoffs are random, being given by the expected payoff in the game $\hat{G}$ plus the outcome R of a random variable $\tilde{R}$. I interpret this random variable as capturing a variety of random shocks that perturb payoffs. I regard this randomness as a crucial feature of many real-world games, where payoffs are likely to be affected by a wide assortment of forces that have been excluded when constructing the model $\hat{G}$. These forces may be sufficiently complicated that players encounter difficulties simply identifying their payoffs precisely, much less determining those payoffs exactly by simply choosing between strategies X and Y. Payoffs are then naturally modeled as being random. The expected value of $\tilde{R}$ is zero, and the distribution of $\tilde{R}$ does not depend upon the strategy chosen by a player or her opponent.[4]

An agent playing X and meeting an opponent who plays X receives an expected payoff of A and an actual payoff of $A + R$. If an agent plays

---

3. Because we shall take $\tau \to 0$, this assumption requires that the game be played arbitrarily rapidly. I view this as an approximation of the case when play is frequent relative to strategy revision, which I consider the natural setting for evolutionary models. Kandori, Mailath, and Rob [128] examine periods of fixed length and assume that either agents play an infinite number of times or agents play a round-robin tournament in each period. Nöldeke and Samuelson [168] assume a round-robin tournament in each period. Young's model [250] is less demanding in this respect, though it still assumes that the result of each game is made known to all agents as soon as it is played. Robson and Vega-Redondo [193] examine periods of fixed length and assume that each agent plays once in each period.

4. For all strategy choices, the distribution F of $\tilde{R}$ is then independent and identically distributed across players. It would be interesting to study cases in which the distribution of $\tilde{R}$ differs across players or strategies, or in which this source of noise is correlated across individuals, perhaps as a result of environmental factors that impose a common risk on all agents. The latter type of uncertainty is discussed by Fudenberg and Harris [85] and by Robson [192].

X in a population where proportion k of her opponents play X, then the agent's expected payoff, denoted by $\pi_X(k)$, is given by

$$\pi_X(k) = kA + (1 - k)C, \tag{3.1}$$

where the expectation is taken with respect to the likely identity of the agent's opponent and the likely realization of $\tilde{R}$. The expected payoff of an agent playing Y in a population where proportion k of her opponents pay X, denoted by $\pi_Y(k)$, is given by

$$\pi_Y(k) = kB + (1 - k)D. \tag{3.2}$$

In each period of length τ, each agent takes a draw from an independently distributed Bernoulli random variable. Let the units in which time is measured be chosen so that with probability τ, this distribution produces the outcome "learn." With the complementary probability, the agent does not learn. An agent who receives the learn draw recalls her realized payoff in the last period and assesses it as being either "satisfactory" or "not satisfactory."[5] If the realized payoff exceeds an aspiration level, then the strategy is deemed satisfactory and the agent makes no change in strategy. If instead the realized payoff falls short of the aspiration level, then the agent loses faith in her current strategy and abandons it. If an agent plays strategy X in a population in which everyone plays X, for example, and then receives the learn draw, the corresponding probability that this agent abandons her strategy is given by

$$g(A) = \mathrm{prob}(A + R < \Delta) = F(\Delta - A), \tag{3.3}$$

where Δ is the aspiration level, R is the realization of the random payoff variable $\tilde{R}$, and F is the cumulative distribution of $\tilde{R}$. Similar expressions hold for the payoffs B, C, and D.

I assume that F is a Uniform distribution on the interval $[-\omega, \omega]$, where $\{A, B, C, D\} \subset [\Delta - \omega, \Delta + \omega]$. Two properties of the Uniform distribution make it especially convenient. First, it satisfies the monotone likelihood ratio property, which is a necessary and sufficient condition for it to be more likely that low average realized payoffs are produced

---

5. Satisficing models have long been a prominent alternative to models of fully rational behavior, being pioneered in economics by Simon [217, 218, 219] and in psychology by Bush and Mosteller [53], and pursued, for example, by Winter [249] and by Nelson and Winter [166]. More recently, satisficing models built on aspiration levels have been examined by Bendor, Mookherjee, and Ray [12] and by Gilboa and Schmeidler [102, 103, 104].

by low average expected payoffs, and hence for realized payoffs to provide a useful basis for evaluating strategies (see Milgrom [157]). If this were not the case, then I would not expect to encounter agents using learning processes in which low-payoff realizations prompt them to seek new strategies.[6] Second, the linearity of the Uniform distribution allows us to pass expectations inside the function g defined in (3.3). Hence, if an agent plays a strategy that gives an expected payoff of π, then the probability she abandons her strategy is given by

$$g(\pi) = F(\Delta - \pi). \tag{3.4}$$

If the agent plays X and proportion k of the opponents play strategy X, then the π that appears in (3.4) is given by $\pi_X(k)$, defined in (3.1). If the agent chooses Y, then π is given by $\pi_Y(k)$, defined in (3.2).

The assumption that F is the Uniform distribution, as well as the strong independence assumption on F, plays an important role in leading to the replicator dynamics. An approximation result similar to proposition 3.1 below could be established for other specifications, but the approximating differential equations would not be the replicator dynamics, which are thus a quite special example of a more general phenomenon. Perhaps the strongest assumption in this model is that the aspiration level is exogenously fixed and does not change over time. Relaxing this assumption introduces a new element of history dependence that qualitatively changes the analysis.[7]

If agent i has abandoned her strategy as unsatisfactory, she must now choose a new strategy. I assume that she randomly selects a member j of the population. With probability $1 - \lambda$, agent i imitates agent j's strategy.[8] With probability λ, agent i is a "mutant" who chooses the strategy that agent j is not playing.

6. I view the learning rule as being the product of an evolutionary process in which agents have discovered that some rules for choosing and rejecting strategies work better than others. The evolutionary process shaping the rules by which agents learn about strategies is likely to proceed at a slower pace than the rate at which agents use these rules to choose strategies, perhaps even being a biological phenomenon, and attention is restricted here to the question of how agents behave given a fixed learning rule. However, if a low payoff does not lead to the expectation that a strategy is likely to yield low payoffs in the future, then evolution is unlikely to produce a learning rule centered around abandoning strategies with low payoffs.

7. Models with endogenous aspiration levels include those of Bendor, Mookherjee, and Ray [12], Börgers and Sarin [48], and Karandikar et al. [130].

8. She may thereby end up playing the strategy with which she began, having perhaps had her faith in it restored by seeing it played by the person she chose to copy.

3.2 Time

How we examine this evolutionary model depends to a large extent on which time period we are interested in. To focus the discussion, let $A > C$ and $B > D$, so that (X, X) and (Y, Y) are Nash equilibria of game $\hat{G}$.

Stationary Distributions

The Aspiration and Imitation model is a Markov model. The state space is given by $\{0, 1, \ldots, N\}$, where a state identifies the number of agents in the population playing strategy X. The transition probabilities, derived from the process by which agents abandon their existing strategies and adopt new ones, will be calculated in the next section.

One approach to the Aspiration and Imitation model, then, is to study its stationary distributions. This study is made easier by the fact that the model has only one stationary distribution. In particular, it is possible for the Aspiration and Imitation model to reach any state from any other state. With the help of just the right combination of learn draws, imitation draws, and mutations, such a transition can always occur in a single step, though it may be most likely to occur through a sequence of steps. It is a standard result that such a Markov process has a unique stationary distribution. The Markov model converges to this distribution from any initial position, in the sense that the distribution describing the likely state of the system at time t approaches the stationary distribution as t gets large. In addition, the distribution of realized states along any sample path approaches the stationary distribution almost surely. Hence the stationary distribution is independent of initial conditions. We can study the outcome of the learning process without having to devote any attention to its point of departure. In the most convenient case, the stationary distribution will concentrate virtually all of its probability on a single equilibrium, which we can then refer to as the outcome or the selected outcome of the stationary distribution.

The models of Kandori, Mailath, and Rob [128] and of Young [250] similarly have the property that for any two states, the probability of a transition from one state to another is positive, and these models accordingly also have unique stationary distributions. Young and Kandori, Mailath, and Rob show that techniques taken from Freidlin and Wentzell [81] can be used to conveniently calculate the stationary distribution. In other models, there may be multiple stationary distributions. We then face the task of characterizing the *set* of stationary distributions, but an evolutionary analysis can still concentrate on these distributions.

The extent to which a stationary distribution is interesting depends upon the length of time required for it to be a good approximation of the behavior of the Markov process. To see the issues involved here, notice that in the Aspiration and Imitation model, the process of rejecting strategies whose payoffs fall short of the aspiration level and imitating the strategies of others will tend to push agents toward best replies. This in turn tends to push the process toward the state in which either all agents play X or all agents play Y, whichever is currently a best reply. Once the system reaches the vicinity of such a state, where all agents play X, for example, it will tend to remain there. Occasionally, the system will move away from the endpoint of the state space as the randomness in payoffs causes agents to abandon X or mutations introduce strategy Y. These departures from the endpoint will usually be small, and will give rise to forces pushing back toward the endpoint. On rare occasions, however, a combination of mutations and realizations of the aspiration and imitation process will switch enough agents to strategy Y as to make the latter a best response, and the expected movement of the system will be toward the state in which all agents play Y. The system is then likely to linger in the vicinity of the state where all agents play Y for a long time, with occasional slight departures, until another coincidental sequence of learning outcomes and mutations restores the state where all agents play X. Similar behavior emerges in Kandori, Mailath, and Rob [128], whose learning process spends most of its time near an endpoint of the state space, being occasionally pushed away by mutations that are usually followed by a movement back to the end of the state space, but which on rare occasions are followed by a switch to the other end of the state space.

In both the Aspiration and Imitation and the Kandori, Mailath, and Rob models, the stationary distribution becomes relevant only after there has been enough time, and enough transitions from one end of the state space to the other, to establish a stable proportion of time spent at each end of the state space. Because the transitions from one end of the state space to the other require a very special combination of events, they can be quite unlikely, and the time required for the stationary distribution to be relevant can be very long.

Although these considerations suggest that the stationary distribution may sometimes refer to too long a time period to be relevant, they also suggest that much can be learned by examining the behavior of the system over periods of time too short for the stationary distribution to be relevant. I refer to such behavior as "sample path" behavior. The time required to reach the stationary distribution can be very long because the sample path is likely to approach one end of the state space and remain in the vicinity of this end, though subjected to frequent, small perturbations, for a long time. But then sample path behavior is likely to remain approximately unchanged over long periods of time. This initial long period of stable behavior may be the most interesting aspect of the learning process and may be the most important target of study.

Sample Paths

It is helpful to organize our thinking about sample paths of learning processes around three aspects of these paths. First, even if the stationary distribution is unique and hence independent of initial conditions, the strategies with which agents happen to begin play will make a difference for the behavior of sample paths. If most agents initially play X in game $\hat{G}$, our expectation is that the learning model will increase the proportion of agents playing X. If most agents initially play Y, then we expect the proportion playing Y to increase. The first case leads to the equilibrium (X, X) being played in most matches for a long period, while the second leads to (Y, Y).

The choices agents make when faced with the task of determining their initial play in a new situation may have little to do with the considerations captured by game theorists when modeling such situations. In particular, I suspect that people seldom have a sufficiently good grasp of the strategic realities, even of situations they encounter frequently, as to analyze these situations in game-theoretic terms.[9] Instead, they simply adopt whatever behavior they typically use or whatever behavior is customary or is the social norm in such situations. If they appear to be optimizing, it is because the social norm has had a

---

9. For example, experiments with second-price auctions reveal that even when people are good at choosing optimal bids, players seldom provide an account of why such a bid is optimal that would be deemed coherent by a game theorist. Similarly, successful poker players are often unable to provide a strategic theory of why bluffing is a good idea.

long time to become adapted to the situations in which it is commonly applied.

We must therefore expect that players faced with a novel situation will choose their action by calling on their experience in similar situations they have faced in the past and then applying the rule of thumb or social norm that guides their behavior in such situations. However, players' assessments of what is similar may not coincide with the judgment of a game theorist, especially as players actually faced with the prospect of making choices in games may find the actual environment much more complicated than the game theorist's corresponding model. The social norm invoked in a novel situation may then be a function of superficial features of the way the problem is posed rather than the strategic aspects of the problem, and a social norm may be triggered that is not at all adapted to the current game. Initial behavior may accordingly be heavily dependent upon the way a new social situation is framed and may bear little relationship to the structure of the problem.

It has become common to offer evolutionary interpretations of the behavior of subjects in laboratory experiments. The evolutionary approach seems natural in light of the observation that subjects' behavior often changes as they gain experience with the game. However, experimental subjects are typically thrust into a quite unfamiliar setting. Their initial behavior may then depend upon a host of details that game theorists would regard as strategically irrelevant. It is thus unsurprising that framing effects are important in experiments and that evolutionary interpretations of experimental results may have little to tell us about initial play.[10]

As experience with a game is accumulated, a player's behavior is likely to adapt to the new circumstances posed by a game. Players begin to learn. This is the second interesting aspect of sample path behavior.

How quickly does learning proceed, and how soon does it lead to seemingly stable behavior? This depends upon three features of the context in which the learning occurs. The first is the extent to which

---

10. To avoid this problem, one might follow Crawford [66] and Roth and Erev [195, 196] in taking the initial behavior as determined by considerations lying outside the model and concentrating on explaining the learning process that appears as this initial behavior gives way to strategic considerations learned in the course of accumulating experience with the game. Other examinations of learning in experimental contexts include Andreoni and Miller [2], Crawford [65], and Miller and Andreoni [158].

the social norm triggered by the framing of the situation has been reinforced in the past and appears to work in the current game. A weakly held norm will be much more easily displaced by learning than a strongly held norm. If different norms are adopted by different players, as may occur when the framing of the problem is relatively abstract, then the subsequent interaction among the players will produce unexpected behavior. None of the norms will seem to work well, prompting reconsideration of the norms that have been adopted and facilitating learning. A norm that is strongly held by all players, however, may be very difficult to dislodge and may cause behavior to respond to learning only very slowly.

Second, learning will be faster the more important the game is to the players. I believe that in their daily lives people face sufficiently many choices to make it impossible for them to analyze all of them. Many, perhaps most, choices are made according to rules or norms that receive occasional scrutiny at best. The full power of our decision-making machinery is reserved only for the most important choices. A choice can fail to be important either because it is rarely made or because the stakes in making the choice are simply too small to bother with. In either case, learning is likely to proceed quite slowly.

Finally, learning will be faster the simpler the task is that the players must learn. It is a common practice for game theorists to assume that players *will* solve any problem that *can* be solved. This may be appropriate if the agents can accumulate enough experience with the game to learn the optimal behavior, but we must recognize that in the case of a difficult task, the length of time required to generate this experience may be very long.

The learning process may lead to a state in which the learning dynamics exert no further pressure. For example, this will happen when the Aspiration and Imitation model or the Kandori, Mailath, and Rob model is applied to Game $\hat{G}$, with the system approaching a state where nearly all players choose X or nearly all players choose Y. Because (X, X) and (Y, Y) are Nash equilibria of Game $\hat{G}$, there is very little incentive for further changes in behavior. The system may then remain in the vicinity of this equilibrium for a very long period of time.

The previous discussion of the stationary distribution shows that the first sojourn in the vicinity of an equilibrium is not the end of the story. The learning dynamics may lead to a state in which they place no more pressure on the system, but there may be other sources of "noise" capable of moving the system between states. This noise may take the form of mutations that perturb a deterministic learning process, as in Kandori, Mailath, and Rob [128], or unexpected realizations of a stochastic learning process such as the Aspiration and Imitation model. Occasionally, these perturbations will be sufficiently severe as to cause the system to jump out of the basin of attraction, under the learning dynamics, of the existing equilibrium and into the basin of attraction of another equilibrium. The learning dynamics will then cause the system to spend a long time near this new equilibrium. Given sufficient time, shocks of this nature will give rise to a distribution describing the relative lengths of time the system spends near each of the possible equilibria. This distribution is a stationary distribution of the system.

How Long Is Long?

It will be helpful to have some terms to describe the various aspects of behavior that potentially emerge from learning models. Binmore and Samuelson [26] describe these aspects of behavior in terms of lengths of time. Like the "short run" and "long run" that are familiar from the theory of the firm, these periods of time are relative terms designed to organize our thinking rather than identify precise lengths of time. They involve not absolute units of time but rather the periods of time it takes certain things to happen. As a result, the length of the various "runs" will vary from setting to setting. Which length is interesting depends upon the game to be played and the setting where it is played.

As do Binmore and Samuelson [26], I shall use "short run" to describe the initial play of the system. I follow both Binmore and Samuelson [26] and Roth and Erev [195] in using "medium run" to describe the behavior of the system as learning begins to have its effect but before the system reaches an equilibrium.

We now might proceed by referring to a stationary distribution of the model as the "long run." This would be consistent with the terms used by Roth and Erev [195] and by Kandori, Mailath, and Rob [128]; it would also be consistent with "long run" as used in economics to mean a period of time long enough for all adjustments to have taken place and as used in mathematics to mean a limiting result as the system is allowed to run arbitrarily long. However, such a usage does not take into account that before a stationary distribution becomes applicable, the sample path of the process may spend a very long time in one place, and the length of time for a stationary distribution to be relevant may be very much longer. In recognition of this, I shall use "long run" to refer to the length of time required for the sample path to reach an equilibrium, near which it will linger for a long time, and "ultralong run" to refer to the length of time required for mutations and other rare events to occur with sufficient frequency to make a stationary distribution relevant.

The issue in determining how to use "long run" is one of modeling. Models are always approximations, and one never literally believes there is nothing else that can happen and that has not been captured by the model. This chapter will show that learning processes, such as the Aspiration and Imitation process or the processes examined by Kandori, Mailath, and Rob [128], are well approximated over potentially very long periods of time by systems of differential equations that miss only certain very improbable events, such as certain very special realizations of the stochastic learning process or large bursts of mutations. A common approach is then to work with such models and to take the resulting differential equations to be an effectively complete specification of the underlying process. Once an equilibrium is reached, there is nothing more *in the model* that can possibly happen. This is commonly called a "long-run analysis" (see Weibull [248], for example), and this is the sense in which I use "long run." In the underlying process, however, there are more things that can and eventually will happen, though they happen with low probability and their occurrence will require waiting a long time. An analysis of a stationary distribution of the process must take account of these possibilities, and the need to characterize such an investigation in terms of lengths of time prompts the introduction of the phrase "ultralong run."

An analogy from the theory of the firm will be useful. The "long run" in the theory of the firm describes the outcome for a fixed specification of the market structure. Over longer periods of time, this long-run outcome will be rendered irrelevant by technological innovations and new product developments that produce large changes in the nature of the market. We know that these will happen, but also expect them to occur so infrequently that a given market equilibrium is likely to persist for a long time. A useful analytical approximation is then to work with models in which market structure is simply fixed, and to refer to the equilibria of these models as "long-run behavior." Someone interested in new product innovations might then object that we have used "long run" to refer to a period of time too short to capture everything that can happen in the market, and that "long run" should be reserved only for behavior emerging from models that allow innovation. At one level, this argument is compelling, but if we are convinced that the resulting lengths of time are too long to be relevant, the most useful approach is to exclude innovation from the model and study the long-run behavior of the resulting model. This is the counterpart of the long-run analysis we shall be undertaking.

This discussion assumes that the sample path of the process will converge to *something*. We shall concentrate on sufficiently simple games that convergence problems do not arise. However, Shapley's celebrated demonstration that fictitious play need not converge in a 3 × 3 normal-form game [216] has since been joined by a number of examples showing that complicated behavior can appear even in seemingly simple games.[11]

3.3   Sample Paths

This section establishes a link between the replicator dynamics and the long-run behavior of the Aspiration and Imitation model. If the population is sufficiently large and the length of a time period is sufficiently short, then the sample path behavior of the Aspiration and Imitation model is approximated arbitrarily closely by a solution of the replicator dynamics. More generally, deterministic differential equations can be good tools for studying the long-run behavior of stochastic selection processes.

The variant of the replicator dynamic to be studied has the form

$$\dot{z} = z(\pi_X(z) - \overline{\pi}(z)) + \lambda(1 - 2z)(K - \overline{\pi}(z)), \tag{3.5}$$

where $z \in [0, 1]$ is interpreted as the fraction of an infinite population playing strategy X in game $\hat{G}$, $\overline{\pi} = z\pi_X(z) + (1 - z)\pi_Y(z)$ is the expected payoff in the population, K is a constant that does not depend on the state, and where

$$\pi_X(z) = zA + (1 - z)C \tag{3.6}$$

$$\pi_Y(z) = zB + (1 - z)D. \tag{3.7}$$

The more familiar form of the replicator dynamic given by (2.15) is obtained by following the standard practice of considering the case in which the probability λ of a mutation is zero.

Equations (3.6)–(3.7) match equations (3.1)–(3.2), with z replacing k. As a result, $\pi_X(z)$ identifies the payoff to an agent playing strategy X

---

11. For example, see Skyrms [220, 221].

given that proportion z of the agent's opponents play X. If the state of a finite population were z, meaning that proportion z of the population plays strategy X, then each individual playing strategy X faces opponents who play X in slightly smaller proportion than z. As the size of the population grows, this distinction disappears and z becomes both the current state and the proportion of opponents playing strategy X.

When the population size is N and the length of a time period is τ, the Aspiration and Imitation model is a Markov process with $N + 1$ population states $x \in \{0, \nu, 2\nu, \ldots, 1\}$, where $\nu = 1/N$ and xN is the number of agents characterized by strategy X in state x. The passage from this model to a differential equation in which z is modeled as taking on values in $[0, 1]$, as in (3.5), will be accomplished by examining the limit as N gets large and τ gets small.

The key transitions in this Markov process are those that lead from an existing state to an adjacent state, meaning a state in which either one more or one fewer agents play strategy X. Because attention will be focussed on short time periods, the probability that two agents receive the learn draw in any single period, and hence that the system potentially moves more than one step in a period, will be negligible. For fixed N, we can calculate the probability $r(x, \nu, \tau)$ of moving one step to the right in a single period, from a state x with $0 \leq x < 1$ to the state $x + \nu$:

$$r(x, \nu, \tau) = \tau N (1 - \tau)^{N-1} (1 - x)\{(x(1 - \lambda) + (1 - x)\lambda)g(\pi_Y(x)) + O(\nu)\}$$
$$+ O(N^2 \tau^2)$$
$$\equiv \tau N (1 - \tau)^{N-1} R(x, \nu) + O(N^2 \tau^2). \tag{3.8}$$

To see how this probability is constructed, notice that for the population to move from state x to state $x + \nu$, in which the number of agents characterized by strategy X has increased by one, four things must happen. First, at least one agent must receive the learn draw. The probability that *exactly* one agent receives the learn draw is $\tau N (1 - \tau)^{N-1}$. The probability that two or more agents receive the learn draw, and that the result is to move the state one step to the right, is $O(N^2 \tau^2)$.[12] Second, the agent receiving the learn draw must currently be playing strategy Y, the probability of which is $1 - x$. Third, the result of this

---

12. The probability that two or more learn draws occur and move the state one step to the right is less than or equal to the probability of two or more learn draws, which is given by $\sum_{k=2}^{N} C_k^N \tau^k (1 - \tau)^{N-k}$, where C_k^N is the relevant binomial coefficient. This is in turn bounded above by $\sum_{k=2}^{N} N^k \tau^k \leq 2N^2 \tau^2 = O(N^2 \tau^2)$, where the inequality requires $N\tau < \frac{1}{2}$, which is ensured by taking the limit $\tau \to 0$ before $N \to \infty$.

agent's strategy assessment must be to abandon strategy Y. An agent playing strategy Y faces opponents who play strategy X in proportion $xN/(N-1)$ and play strategy Y in proportion $(N - xN - 1)/(N-1)$. The probability of abandoning Y is then given by $g(\pi_Y(x) + O(v)) = g(\pi_Y(x)) + O(v)$, where this equality follows from (3.3) and from F being the Uniform distribution.[13] Finally, when choosing a new strategy, the agent must choose strategy X, either by correctly imitating an agent playing X or by mistakenly imitating an agent playing Y. The probability of choosing X is given by

$$\frac{xN}{N-1}(1-\lambda) + \lambda\frac{N-xN-1}{N-1} = x(1-\lambda) + (1-x)\lambda + O(v).$$

Similarly, the probability $\ell(x, v, \tau)$ of moving one step to the left in a single period from a state x with $0 < x \le 1$ to the state $x - v$ is given by

$$\ell(x, v, \tau) = \tau N(1-\tau)^{N-1}x\{((1-x)(1-\lambda) + x\lambda)g(\pi_X(x)) + O(v)\}$$
$$\quad + O(N^2\tau^2)$$
$$= \tau N(1-\tau)^{N-1}L(x, v) + O(N^2\tau^2). \tag{3.9}$$

Note that (3.8) and (3.9) implicitly define $R(x, v)$ and $L(x, v)$ by

$$R(x, v) = (1-x)\{(x(1-\lambda) + (1-x)\lambda)g(\pi_Y(x)) + O(v)\} \tag{3.10}$$
$$L(x, v) = x\{((1-x)(1-\lambda) + x\lambda)g(\pi_X(x)) + O(v)\}. \tag{3.11}$$

These probabilities of moving one step to the left or right can now be used to construct an approximate characterization of sample path behavior. The first step is to estimate the expectation and variance of $x(t + \tau) - x(t)$, conditional on $x(t) = x$. To this end, we can introduce the quantities

$$\mu(x, v) = R(x, v) - L(x, v)$$
$$= x(1-x)(g(\pi_Y) - g(\pi_X)) + \lambda(1-2x)\bar{g} + O(v) \tag{3.12}$$
$$\sigma^2(x, v) = R(x, v) + L(x, v)$$
$$= (1-2\lambda)x(1-x)(g(\pi_Y) + g(\pi_X)) + \lambda\bar{g} + O(v), \tag{3.13}$$

where $\bar{g} = xg(\pi_X(x)) + (1-x)g(\pi_Y(x))$.

It is useful to note that (3.12)–(3.13) allow us to write the replicator equation (3.5) in the form

$$\dot{z} = \mu(z, 0). \tag{3.14}$$

---

13. This uses the fact that $N/(N-1) = 1 + 1/(N-1)$ and $1/(N-1) = 1/N + 1/(N(N-1))$.

To obtain the result in (3.14), we recall that F is a Uniform distribution on $[-\omega, \omega]$, and hence $g(\pi) = (\Delta + \omega - \pi)/2\omega$, and then calculate

$$\mu(x, 0) = x(1 - x)\{g(\pi_Y(x)) - g(\pi_X(x))\}$$
$$+ \lambda(1 - 2x)\{(1 - x)g(\pi_Y(x)) + xg(\pi_X(x))\}$$
$$= \frac{x(1 - x)}{2\omega}(\pi_X(x) - \pi_Y(x)) + \frac{\lambda(1 - 2x)}{2\omega}(\Delta + \omega - \overline{\pi}). \quad (3.15)$$

Letting $K = \Delta + \omega$, rescaling time to eliminate the 2ω in the denominators, noting that $x(1 - x)(\pi_X(x) - \pi_Y(x)) = x(\pi_X(x) - \overline{\pi}(x))$, and replacing the discrete variable x with the continuous variable z then gives (3.14).

Returning to the estimation of the expectation and variance of $x(t + \tau) - x(t)$, let $x(t + \tau) - x(t) \equiv \delta x(t)$. Then, since $\mathcal{E}\{x(t)|x(t) = x\} = x$ and the probability of moving more than one step in a single period is $O(N^2\tau^2)$, (3.12) and (3.13) lead to the following:

$$\mathcal{E}\{\delta x(t)|x(t) = x\} = vr(x, v, \tau) - v\ell(x, v, \tau) + O(N^2\tau^2)$$
$$= \tau\mu(x, v) + O(N^2\tau^2), \quad (3.16)$$
$$\mathcal{E}\{(\delta x(t))^2|x(t) = x\} = v^2r(x, v, \tau) + v^2\ell(x, v, \tau) + O(N^2\tau^2)$$
$$= \tau v\sigma^2(x, v) + O(N^2\tau^2), \quad (3.17)$$
$$\text{var}\{\delta x(t)|x(t) = x\} = \tau v\sigma^2(x, v) - \{\tau\mu(x, v)\}^2 + O(N^2\tau^2)$$
$$= \tau v\sigma^2(x, v) + O(N^2\tau^2). \quad (3.18)$$

Rewrite equation (3.16) as

$$\frac{\mathcal{E}\{x(t + \tau)|x(t) = x\} - \mathcal{E}\{x(t)|x(t) = x\}}{\tau} = \mu(x, v) + O(N^2\tau). \quad (3.19)$$

When $(\tau, N) \to (0, \infty)$ in (3.19) so that $N^2\tau \to 0$, the right side converges to $\mu(x, 0)$. Notice that two limits are involved here, namely, $\tau \to 0$ and $N \to \infty$, and that the relative rates at which these limits are taken are not arbitrary. Instead, τ must approach zero sufficiently fast that $N^2\tau$ approaches zero. If one limit is taken after the other, the order must be $\tau \to 0$ first, and then $N \to \infty$. I return to the order in which various limits are taken in section 3.5.

When $\tau \to 0$ and then $N \to \infty$, the left side of (3.19) converges to a derivative that informal derivations of the replicator dynamics often identify with $x'(t)$ without further comment. If this identification is appropriate, (3.19) would produce the replicator equation in the limit. Börgers and Sarin [47] point out that some care must be taken in performing this limiting operation and that a formal argument is necessary. The proof of the follc*ving proposition is constructed along the lines of Boylan [49, 51].[14] In this proposition, it is important to distinguish between the state $x(t)$ of the system described by the Aspiration and Imitation model and the state $z(t)$ of the system governed by the replicator equation (3.14). The former is a random variable defined on the discrete set of points $\{0, \tau, 2\tau, \ldots\}$, while the latter is the deterministic solution to a differential equation whose state space is the interval $[0, 1]$. Section 3.6 proves

Proposition 3.1 Fix $x(0)$ and let $z(t)$ be the solution of the replicator equation $\dot{z} = \mu(z, 0)$ subject to the boundary condition $z(0) = x(0)$. Then for any $\epsilon > 0$ and any integer $T > 0$, there are positive constants N_0 and τ_0 such that if $N > N_0$ and $N^2\tau < \tau_0$, then

$$\text{prob}\{|z(t) - x(t)| \geq \epsilon\} < \epsilon,$$

for any t satisfying $0 \leq t \leq T$ at which $x(t)$ is defined.

Because proposition 3.1 holds for finite values of t, it is not a characterization of the stationary distribution of the Aspiration and Imitation Markov process. Instead, it establishes properties of the sample paths of the model. I refer to these sample path properties as "having long-run implications." To see what is meant by this and to explore the implications, suppose that $z(t)$ is governed by the replicator dynamic (3.5) (or (3.14)) with the boundary condition $z(0) = x(0)$. Figure 3.2 shows a phase diagram for the replicator dynamic (3.5) when λ is small and when $A > B$ and $D > C$, so there are two strict Nash equilibria, given by (X, X) and (Y, Y), and one mixed-strategy equilibrium. There are three stationary states ξ, κ and ζ satisfying $0 < \xi < \kappa < \zeta < 1$. The inner stationary state is unstable, but ξ and ζ are asymptotically stable with basins of attraction $[0, \kappa)$ and $(\kappa, 1]$ respectively. As $\lambda \to 0$, the stationary states ξ, κ, and ζ converge to the three Nash equilibria of game $\hat{G}$. The stationary state ξ converges to the equilibrium in which all agents play Y. The stationary state ζ converges to the equilibrium in which all agents play X. The stationary state κ converges to the mixed equilibrium.

---

14. I thank Rob Seymour for his help in understanding these methods. Börgers and Sarin [47] establish a similar result by appealing to theorem 1.1 of Norman [170, p. 118], deriving the replicator dynamics in a model in which the "population size" is infinite from the outset.

Figure 3.2
Phase diagram for game $\hat{G}$

Suppose that $z(0) = x(0)$ happens to lie in the basin of attraction of
ζ. The replicator dynamics will then move close to ζ and remain in the
vicinity of ζ forever. For sufficiently large N and small τ, we have from
proposition 3.1 that $|z(t) - x(t)| < \epsilon$ with probability at least $1 - \epsilon$ for
$t \leq T$. With high probability, the Aspiration and Imitation model thus
also moves close to ζ and remains in the vicinity of ζ for an extended
time. Upon observing such behavior, we would be tempted to say that
the process has converged to ζ or has "selected" the equilibrium ζ. This
long period of time spent in the vicinity of a state such as ζ qualifies
this as a long-run result.

In examining limiting behavior, the limit $t \to \infty$ was applied to the
process $z(t)$ and *not* to the process $x(t)$. As the next section on the
ultralong-run behavior of $x(t)$ demonstrates, the ultralong-run behavior of the Aspiration and Imitation model may be quite different from
that described by the replicator dynamics in the long run. The larger
is the population size N, the longer will the approximation provided
by the replicator dynamics be a good one, but the size of the population required for this approximation increases as the interval of time
increases, and the replicator dynamics will not be a good approximation forever.

3.4   Stationary Distributions

This section studies the stationary distribution over states given by the
Aspiration and Imitation model. In the terms given above, this is an
ultralong-run analysis.

Again, the analysis begins with the observation that for fixed values
of τ, ν and λ, the Aspiration and Imitation model is a Markov process
on the state space $\{0, \nu, 2\nu, \ldots, 1\}$ with a single ergodic set consisting of
the entire state space. Hence the following standard result for Markov
processes is immediate (see Billingsley [17], Seneta [214], or Kemeny
and Snell [132, theorems 4.1.4, 4.1.6, and 4.2.1]):

Proposition 3.2   Fix τ, v, and λ. Then there exists a unique probability distribution $P(x, v, \tau)$ on $\{0, v, 2v, \ldots, 1\}$ that is a stationary distribution for the Aspiration and Imitation Markov process. The relative frequencies of the various states realized along a sample path approach the distribution $P(x, v, \tau)$ almost surely as $t \to \infty$. For any initial condition $x(0)$, the probability distribution describing the likely state of the process at time t converges to $P(x, v, \tau)$ as t gets large.

The distribution $P(x, v, \tau)$ is referred to as either the "stationary distribution" or the "asymptotic distribution" of the Markov process, where $P(x, v, \tau)$ gives the probability attached to state x given v and τ.

Let us now turn to characterizing the stationary distribution. For the case of small values of τ, we can derive a particularly convenient representation for $P(x, v, \tau)$. Let $P(x, v, \tau, t)$ be the probability attached to state x by the Markov process at time t, given v and τ. Then for small values of τ, this satisfies

$$P(x, v, \tau, t+\tau) = P(x + v, v, \tau, t)\ell(x + v, v, \tau)$$
$$+ P(x - v, v, \tau, t)r(x - v, v, \tau)$$
$$+ P(x, v, \tau, t)\{1 - \ell(x, v, \tau) - r(x, v, \tau)\}$$
$$+ O(N^2\tau^2), \tag{3.20}$$

where $\ell(x, v, \tau)$ and $r(x, v, \tau)$ are defined by (3.8)–(3.9), where $\ell(0, v, \tau) = r(1, v, \tau) = 0$, and where $P(1 + v, v, \tau, t) = P(-v, v, \tau, t) = 0$. The error term $O(N^2\tau^2)$ in (3.20) arises from the possibility that there may be multiple recipients of the learn draw in a single period. The stationary distribution $P(x, v, \tau)$ is then characterized by the following version of (3.20):

$$P(x, v, \tau) = P(x + v, v, \tau)\ell(x + v, v, \tau) + P(x - v, v, \tau)r(x - v, v, \tau)$$
$$+ P(x, v, \tau)\{1 - \ell(x, v, \tau) - r(x, v, \tau)\} + O(N^2\tau^2). \tag{3.21}$$

Rearranging (3.21) gives

$$0 = P(x + v, v, \tau)\ell(x + v, v) - P(x, v, \tau)\ell(x, v, \tau)$$
$$+ P(x - v, v, \tau)r(x - v, v, \tau) - P(x, v, \tau)r(x, v, \tau) + O(N^2\tau^2). \tag{3.22}$$

Now take the limit $\tau \to 0$. As $\tau \to 0$, the probability of more than one agent receiving the learn draw in a single period becomes negligible. Transitions then consist of either a single step to the left or a single step to the right. The result is a *birth–death chain* (cf. Gardiner [97] and Karlin and Taylor [131, chapter 4]). The following description of the stationary

|→|→|→|→|→|→|→|→|→|→|→|→|→|←|←|←|←|←|←|←|

0 x 1

Figure 3.3
Significant x-tree

distribution is standard for such chains. In particular, (3.23) follows immediately from (3.22), (3.10)–(3.11), and $\ell(0, v, \tau) = r(1, v, \tau) = 0$.

Proposition 3.3   Let $P(x, v) = lim_{\tau \to 0} P(x, v, \tau)$. Then for any state $x \in \{0, v, 2v, \ldots, 1\}$, we have

$$\frac{P(x + v, v)}{P(x, v)} = \frac{R(x, v)}{L(x + v, v)}. \tag{3.23}$$

An alternative route to this conclusion, which highlights the nature of the limiting argument as $\tau \to 0$, uses the method of Freidlin and Wentzell [81], as employed by Young [250] and by Kandori, Mailath, and Rob [128]. One first constructs, for each state x in the finite Markov process, the collection of all x-trees, where an x-tree is a collection of transitions between states with the property that each state other than x is the origin of one and only one transition, x is the origin of no transition, and there is a path of transitions to x from every state other than x. Fixing a state x, one then computes the products of the transition probabilities in each x-tree and adds these products (for all of the x-trees) to obtain a number $Z(x)$. The exact ultralong-run probabilities $P(x, v)$ and $P(x', v)$ of being at states x and x' are then proportional to $Z(x)$ and $Z(x')$. In the Aspiration and Imitation model, products of order greater than τ^N can be neglected because the limit $\tau \to 0$ is to be taken.[15] Only the single x-tree of order $O(\tau^N)$ illustrated in figure 3.3 then need be retained. When the products for x and $x + v$ are divided, only one factor from each fails to cancel, leaving (3.23).

Yet another path to this conclusion is provided by the following. First, rewrite equation (3.20) in the form

$$v \left\{ \frac{P(x, v, \tau, t + \tau) - P(x, v, \tau, t)}{\tau(1 - \tau)^{N-1}} \right\} = P(x + v, v, \tau, t) L(x + v, v)$$

$$- P(x, v, \tau, t) L(x, v) + P(x - v, v, \tau, t) R(x - v, v)$$

$$- P(x, v, \tau, t) R(x, v) + O(N^2 \tau^2).$$

15. For similar reasons, Young [250] and Kandori, Mailath, and Rob [128] can neglect products that do not minimize the order of λ because the limit $\lambda \to 0$ is taken.

Now take the limit $\tau \to 0$. The error term on the right disappears and we are left with the probability $P(x, v, t)$ that satisfies

$$v\frac{\partial P(x, v, t)}{\partial t} = P(x + v, v, t)L(x + v, v) - P(x, v, t)L(x, v)$$

$$+ P(x - v, v, t)R(x - v, v) - P(x, v, t)R(x, v). \qquad (3.24)$$

A stationary distribution is defined by the property that $\partial P(x, v, t)/\partial t = 0$. Writing $\partial P(x, v, t)/\partial t = 0$ in (3.24) leads us to equation (3.22) with a zero error term. Proposition 3.3 therefore still applies.

What is the relationship between the replicator approximation of the Aspiration and Imitation model and the asymptotic distribution? To bring this question into sharper focus, suppose that the initial condition lies in the basin of attraction of ζ relative to the replicator dynamics (see figure 3.2). As a result, the system will initially be highly likely to approach ζ and spend an extended period of time in the neighborhood of ζ. The larger N gets, the more likely this will be a good description of the behavior of the system and the longer the time span over which it will be valid.

Chapter 9 shows that, in spite of this initial behavior, there are payoff specifications for which the asymptotic distribution attaches almost all of its probability mass on ξ. To reconcile these observations, we notice that a transition from a state near ζ to a state in the basin of attraction of ξ requires a rather special realization of random events. These events require a much larger proportion of X agents to abandon their strategies and adopt Y than the system is expected to produce. During any bounded time interval, the probability of such a combination of strategy revisions, and hence the probability that the Aspiration and Imitation population will reach the basin of attraction of ξ, is arbitrarily small. However, "arbitrarily small" is not the same as "zero." The probability that such a combination of strategy revisions will *eventually* occur is unity. As a result, the system will, with certainty, visit states near ξ and spend most of its time in such states, though for some initial conditions and population sizes one will have to wait an extraordinarily long time for this to occur.

3.5   Limits

The taking of limits plays an important role in many evolutionary analyses. The Aspiration and Imitation model is no exception. In the course of examining the model, two limits have been taken: the length of a time period τ has been allowed to approach zero and the population size N has approached infinity. We may also be interested in allowing the mutation probability λ to approach zero, as do Young [250] and Kandori, Mailath, and Rob [128].

The limit $\tau \to 0$ allows the convenience of working in continuous time. In particular, it allows us to ignore the possibility of more than one agent revising a strategy at once, so that strategy assessments are isolated, idiosyncratic events. This will not always be appropriate. In a biological context, reproduction may be coordinated by breeding seasons, and hence may be best captured by a discrete-time model with simultaneous reproduction. In a learning context, the arrival of information or other public events may coordinate strategy assessments. In other cases, however, the times when strategies are revised will not be coordinated, leading to a model in which we can usefully let τ approach zero.

The limit $N \to \infty$ allows us to concentrate on the expected value of the process, at least over finite periods of time. This is typical of evolutionary analyses, which commonly appeal to large-population assumptions not only to work with expected values but also to eliminate repeated-game effects.

What about the order in which the limits $\tau \to 0$ and $N \to \infty$ are taken? Here we shall take the limit $\tau \to \infty$ first, to reflect our interest in an overlapping generations model where strategy revisions occur at uncoordinated times.

After $\tau \to 0$, we have the limits $N \to \infty$ and possibly $\lambda \to 0$. In many cases, I think it is inappropriate to take the limit $\lambda \to 0$. Instead, I believe that the world is an inherently noisy environment, and the most realistic model retains mutations. If we are to take $\lambda \to 0$, however, I believe that this limit should be taken last. In some cases, the order of these limits does not matter. Section 9.4 shows that in 2×2 games with two strict Nash equilibria, reversing the order of limits has no effect and the order is arbitrary. This is not always the case, however. The order matters in 2×2 games with a single, mixed-strategy Nash equilibrium.

To illustrate this point, consider the game of Chicken shown in figure 3.4. The phase diagram under either the replicator or best-reply dynamics is shown in figure 3.5. This game is a special case of the Hawk-Dove Game of Maynard Smith [149], prompting the strategy labels of D (for "Dove") and H (for "Hawk"). Chicken has a unique symmetric equilibrium in which each player plays D and H with

	D	H
D	1, 1	0, 2
H	2, 0	−1, −1

Figure 3.4
Chicken

0
(All H)

1
(All D)

Figure 3.5
Phase diagram for Chicken

equal probability. The replicator dynamics lead to this equilibrium. The stationary distribution of the Aspiration and Imitation model concentrates all of its probability near this equilibrium, given that $N \to \infty$ *before* $\lambda \to 0$ or, equivalently, given that the population is sufficiently large and the mutation rate not too small.[16]

Suppose that the order of the limits $N \to \infty$ and $\lambda \to 0$ is reversed. Before the limit $N \to \infty$ is taken, the system will then have a positive probability of reaching any state x from any initial state $x(0)$ satisfying $0 < x(0) < 1$. But if $\lambda = 0$, the boundary states $x = 0$ and $x = 1$, in which the population consists either of all Hs or all Ds, are absorbing states. Once the population enters such a state, it cannot leave without a mutation. The limit of the stationary distribution, as $\lambda \to 0$, will then concentrate all of its probability on these two nonequilibrium states, with all agents playing a payoff-inferior strategy. This limit persists as $N \to \infty$. The outcome of the system is then determined by accidental extinctions from which the population cannot recover. This contrasts with the selection of the mixed equilibrium when $N \to \infty$ before $\lambda \to 0$, which strikes me as the more appropriate ultralong-run model. In particular, imitation may be important in shaping behavior, but people do sometimes experiment with new strategies and ideas. An outcome that depends crucially on the lack of such experimentation, such as a

---

16. The two asymmetric pure Nash equilibria are not candidates for selection because the Aspiration and Imitation model has only one population from which both players are always drawn. To select an asymmetric equilibrium, it would be necessary to draw the two players from different populations that evolve separately.

stationary distribution concentrated on the endpoints of the state space, should be viewed with suspicion. As a result, if one is interested in large populations and small mutation rates, the limit $\lambda \to 0$ should be taken last (or not taken at all).

This fixes the order of limits as first $\tau \to 0$, next $N \to \infty$, and then $\lambda \to 0$. But when, in the course of taking these limits, do we examine the limiting outcome of the system? The distinction between the long run and the ultralong run hinges on the answer to this question.

The key issue here is whether we examine the system before or after letting the population size approach infinity. First, we might examine the limiting outcome of the system after letting $N \to \infty$. Letting "$t \to \infty$" be a shorthand representation of "examine the limiting outcome of the system," we then have two possibilities:

Case 1: First $\tau \to 0$, next $N \to \infty$, next $\lambda \to 0$, then $\mathbf{t} \to \infty$.

Case 2: First $\tau \to 0$, next $N \to \infty$, next $\mathbf{t} \to \infty$, then $\lambda \to 0$.

These two cases correspond to a long-run analysis. In particular, by first letting time periods get short and then the population get large,[17] we obtain a deterministic differential equation. In the Aspiration and Imitation Model, this equation is a version of the classical replicator dynamics (with an extra term added to take account of the existence of mutations). After obtaining this equation, we let $t \to \infty$, which is a shorthand for the statement that we study the limiting behavior of the *differential equation*.[18] The replicator dynamics are commonly used to examine the limiting outcome of the system with the simplifying assumption that $\lambda = 0$, giving case 1. Case 2 obtains if we retain mutations when studying the limiting outcome of the replicator dynamics.

We must interpret the notation $t \to \infty$ carefully. By assumption, this denotes the operation of examining a limiting outcome, which is ac-

---

17. From proposition 3.1, the precise operation here is $(\tau, N) \to (0, \infty)$, so that $N^2 \tau \to 0$.
18. Cases 1 and 2 can be contrasted with the traditional story (outlined in section 2.3) that is associated with informal motivations of the replicator dynamics: first $\lambda \to 0$, next $N \to \infty$, next $\tau \to 0$, then $t \to \infty$. In particular, by first dispensing with mutations ($\lambda \to 0$) and then letting the population size get large ($N \to \infty$), a model is obtained in which large numbers of simultaneous births occur at discrete intervals. This model is described by a deterministic difference equation known as the "discrete replicator dynamic." This difference equation is commonly approximated by a differential equation ($\tau \to 0$). The limiting behavior of the differential equation is then studied ($t \to \infty$).

complished by literally taking the limit $t \to \infty$ in the replicator dynamics. However, this does not correspond to taking the limit $t \to \infty$ in the underlying model. Instead, we have seen that the replicator dynamics provide a good guide to the behavior of the underlying model only for finite, though possibly very long, time periods. Taking the limit $t \to \infty$ in the replicator dynamics is then a convenient way of approximating the behavior of the Aspiration and Imitation over long, but not infinitely long, periods of time.

The second possibility is that we examine the limiting outcome of the system before taking the limit $N \to \infty$. There are two possibilities here, depending upon whether we take the limit $t \to \infty$ before or after allowing the length of a time period to approach zero, yielding two cases:

Case 3: First $\tau \to 0$, next $t \to \infty$, next $N \to \infty$, then $\lambda \to 0$.

Case 4: First $t \to \infty$, next $\tau \to 0$, next $N \to \infty$, then $\lambda \to 0$.

Cases 3 and 4 correspond to an analysis of the stationary distribution. Here, the first step is to derive the stationary distribution of the discrete-time (case 4) or continuous-time (case 3) Markov process for fixed values of the parameters. We have seen that the order of the limits $t \to \infty$ and $\tau \to 0$ is immaterial in the Aspiration and Imitation Model, and hence these two approaches yield identical results. In particular, we examine this stationary distribution before limiting arguments involving the population size or mutation rate obscure any of the randomness of the Markov process. The remaining limits are comparative static exercises that allow the derivation of particularly sharp approximations of the limiting distribution.

Which of these two approaches, which can give quite different outcomes, is the appropriate one? Should we be interested in the long run or the ultralong run? The easy way out of this question is to claim that both can be interesting, and subsequent chapters are devoted to both. The important point here is that the replicator dynamics and the stochastic models of Young [250] and of Kandori, Mailath, and Rob [128] need not be viewed as competing models or methods, despite their potentially differing outcomes. Instead, they are alternative approximations for examining the outcome of a single model, with one approximation being appropriate when addressing the long run and the other, when addressing the ultralong run.

The random influences that have been lumped together in this chapter under the heading of "mutations" can play a crucial role in determining the outcome of the system. This is most obviously the case in the ultralong run, where the relative probabilities of various combinations of mutations drive the equilibrium selection results of Kandori, Mailath, and Rob [128] and of Young [250]. In ultralong-run models, these are truly random forces, and I shall refer to them as "noise." In the long-run models that emerge from the analysis of cases 1 and 2, these random forces have been "smoothed out" into deterministic flows and appear as the final term in (3.5). I refer to these deterministic flows as "drift." Chapter 6 shows that even when arbitrarily weak, drift can play a crucial role in shaping the outcome of the system.

3.6 Appendix: Proofs

Proof of Proposition 3.1   Fix a value of $T > 0$. Unless otherwise stated, t will be assumed to be admissible, that is, to be of the form $t = k\tau$ for some integer k. Bounds that are not given explicitly hold for all admissible t satisfying $0 \leq t \leq T$.

The first step is to observe that because $\mu(z, v)$ satisfies appropriate Lipschitz conditions, we can replace $z(t)$ by $Z(t)$, where $z(t)$ is the solution to $\dot{z} = \mu(z, 0)$ and $Z(t)$ is the solution of $\dot{Z} = \mu(Z, v)$, each subject to the boundary condition $Z(0) = x(0)$. In particular, $|Z(t) - z(t)| < \frac{1}{2}\epsilon$ provided that N is sufficiently large.[19] It then remains to establish the proposition with z replaced by Z and ϵ replaced by $\frac{1}{2}\epsilon$.

On integrating, we obtain the following expression for $Z(t)$:

$$Z(t) - x(0) = \int_0^t \mu(Z(s), v)\, ds\,. \tag{3.25}$$

The next step is to find a corresponding expression for $x(t) - x(0)$. To this end, we can follow Boylan [51, p. 16] in defining

$$m(k\tau) = x(k\tau) - x(0) - \sum_{j=1}^{k} \mathcal{E}\{x(j\tau) - x(j\tau - \tau)|x(j\tau - \tau)\}\,. \tag{3.26}$$

---

19. From (3.12), there is a $C > 0$ such that for all z and z' in $[0, 1]$, $|\mu(z, 0) - \mu(z', 0)| \leq C|z - z'|$ and $|\mu(z, v) - \mu(z, 0)| \leq Cv$. It is then standard to observe that $|z(t) - Z(t)| \leq \int_0^t |\mu(z(s), 0) - \mu(Z(s), v)|\, ds \leq C\{vt + \int_0^t |z(s) - Z(s)|\, ds\}$. It follows from Gronwall's lemma (Revuz and Yor [187, p. 499]) that $|z(t) - Z(t)| \leq Cvte^{Ct} < \epsilon/2$, provided that N is sufficiently large.

Boylan notes that $m(k\tau)$ is a martingale, which is a fact that will be needed later. For the moment, it is important to note only that $\mathcal{E}\{m(k\tau)\} = \mathcal{E}\{m(0)\} = m(0) = 0$.

On rearranging (3.26) and making use of (3.16), we obtain

$$x(t) - x(0) = m(t) + \sum_{j=1}^{k} \mathcal{E}\{x(j\tau) - x(j\tau - \tau) | x(j\tau - \tau)\}$$

$$= m(t) + \sum_{j=1}^{k} \mu(x(j\tau - \tau), v) + O(N^2\tau^2 k)$$

$$= m(t) + \int_0^t \mu(x([s/\tau]\tau), v)\,ds + O(N^2\tau), \qquad (3.27)$$

where $[s]$ is the integer part of s.

The next step is to subtract (3.27) from (3.25) to obtain

$$|Z(t) - x(t)| \leq |m(t)| + \int_0^t |\mu(Z(s), v) - \mu(x([s/\tau]\tau), v)|\,ds + O(N^2\tau)$$

$$\leq |m(t)| + C\int_0^t |Z(s) - x([s/\tau]\tau)|\,ds + O(N^2\tau), \qquad (3.28)$$

where C has been chosen independently of v so that $|\mu(a, v) - \mu(b, v)| \leq C|a - b|$. Since $t = k\tau$, we have $t = [t/\tau]\tau$. Hence, on writing

$$\phi(s) = |Z(s) - x([s/\tau]\tau)|,$$

inequality (3.28) implies that

$$\phi(t) \leq M(t) + C\int_0^t \phi(s)\,ds, \qquad (3.29)$$

where

$$M(t) = \sup_{0 \leq s \leq t} |m([s/\tau]\tau)| + O(N^2\tau) + O(v).$$

The final error term $O(v)$ has been added to the expression for $M(t)$, so that (3.29) holds for all t satisfying $0 \leq t \leq T$ and not just t of the form $k\tau$. The supremum in the expression for $M(t)$ ensures that M increases. We can then appeal directly to Gronwall's lemma, from which we deduce that

$$\phi(t) \leq M(t)e^{Ct}$$

for all t satisfying $0 \leq t \leq T$.

Restricting ourselves again to the case where time variables take only admissible values of the form $k\tau$, we can recall that $\mathcal{E}\{m(t)\} = 0$ and use Chebychev's inequality to obtain

$$\text{prob}\{|Z(t) - x(t)| \geq \tfrac{1}{2}\epsilon\} \leq \text{prob}\{M(t) \geq \tfrac{1}{2}\epsilon e^{-Ct}\}$$

$$\leq t \max_{0 \leq s \leq t} \text{prob}\{|m(s)| \geq \tfrac{1}{4}\epsilon e^{-Ct}\} \tag{3.30}$$

$$\leq T \max_{0 \leq s \leq T} \text{prob}\{|m(s)| \geq \tfrac{1}{4}\epsilon e^{-CT}\}$$

$$\leq T \max_{0 \leq s \leq T} \frac{16e^{2CT}}{\epsilon^2} \text{var}\{m(s)\}$$

$$\leq T \max_{0 \leq s \leq T} \frac{16e^{2CT}}{\epsilon^2} \mathcal{E}\{m(s)\}^2, \tag{3.31}$$

provided that N is sufficiently large and $N^2\tau$ is sufficiently small.[20]

To make further progress, it is necessary to estimate $\mathcal{E}\{m(t)\}^2$. In so doing, I follow Boylan [51] closely in exploiting the fact that $m(k\tau)$ is a martingale. Define

$$\Delta_i = m(i\tau) - m(i\tau - \tau).$$

Then, if $i > j$, we have

$$\mathcal{E}\{\Delta_i \Delta_j\} = \mathcal{E}\{\mathcal{E}\{\Delta_i \Delta_j | x(j\tau)\}\} = \mathcal{E}\{\Delta_j \mathcal{E}\{\Delta_i | x(j\tau)\}\} = 0,$$

because $\mathcal{E}\{\Delta_i | x(j\tau)\} = \mathcal{E}\{m(i\tau) | x(j\tau)\} - \mathcal{E}\{m(i\tau - \tau) | x(j\tau)\} = m(j\tau) - m(j\tau) = 0$. It follows that

$$\mathcal{E}\{m(k\tau)\}^2 = \mathcal{E}\{\sum_{i=1}^{k} \Delta_i\}^2 = \mathcal{E}\{\sum_{i=1}^{k} \sum_{j=1}^{k} \Delta_i \Delta_j\} = \sum_{i=1}^{k} \mathcal{E}\{\Delta_i^2\}. \tag{3.32}$$

To estimate $\mathcal{E}\{\Delta_i^2\}$, I use (3.18). Since $\mathcal{E}\{\Delta_i\} = \mathcal{E}\{m(i\tau)\} - \mathcal{E}\{m(i\tau - \tau)\} = 0$, we have

$$\mathcal{E}\{\Delta_i^2\} = \text{var}\Delta_i = \text{var}\{x(i\tau) - x(i\tau - \tau) | x(i\tau - \tau)\}$$

$$= \tau v \sigma^2(x(i\tau - \tau), v) + O(N^2\tau^2)$$

$$\leq \tau v S + O(N^2\tau^2), \tag{3.33}$$

---

20. To obtain inequality (3.30), we need N large enough and τ small enough to ensure that the error term $O(N^2\tau) + O(v)$ in the definition of $M(t)$ is less than $\epsilon e^{-Ct}/4$. Inequality (3.30) then follows from the previous inequality by noting that the probability of the largest of t random variables exceeding a constant is no greater than t times the probability of the random variable most likely to do so.

where S is an upper bound for $\sigma^2(x, v)$ that is independent of x and v. Inserting the estimate (3.33) into (3.32), we obtain

$$\mathcal{E}\{m(k\tau)\}^2 \le k\tau vS + O(N^2\tau^2 k) = \frac{tS}{N} + O(N^2\tau) \le \frac{\epsilon^3}{16T}e^{-2CT} \tag{3.34}$$

provided that N is sufficiently large and $N^2\tau$ is sufficiently small.

On using the estimate (3.34) in (3.31), we find that

$$\text{prob}\{|z(t) - x(t)| \ge \epsilon\} \le \text{prob}\{|Z(t) - x(t)| \ge \tfrac{1}{2}\epsilon\}$$

$$\le T \max_{0 \le s \le T} \frac{16e^{2CT}}{\epsilon^2} \frac{\epsilon^3}{16T}e^{-2CT} = \epsilon,$$

which yields the conclusion of the proposition.   □

4   The Dynamics of Sample Paths

Chapter 3 showed that if the population is sufficiently large, then a good approximation of the behavior of the Aspiration and Imitation model, over finite periods of time, can be obtained by examining the deterministic differential equation describing the expected behavior of the system. In the particular case of the Aspiration and Imitation model, the relevant differential equation is a version of the replicator dynamics. A similar approximation result holds for other models, though these will typically give rise to other differential equations. This chapter is devoted to the study of evolutionary models built around systems of differential equations.[1]

4.1   Dynamics

Attention is restricted to two-player, asymmetric games. Let the players be named I and II. The notation easily becomes burdensome, and it will be convenient to let the finite sets of pure strategies for player I and II be denoted by $\mathcal{I}$ and $\mathcal{J}$. An element of $\mathcal{I}$ will typically be denoted by i and an element of $\mathcal{J}$ by j. Payoff functions are then given by $\pi_I : \mathcal{I} \times \mathcal{J} \to \mathbb{R}$ and $\pi_{II} : \mathcal{I} \times \mathcal{J} \to \mathbb{R}$. Let the number of strategies in $\mathcal{I}$ and $\mathcal{J}$ be denoted n_I and n_{II}. Let x and y be elements of S^{n_I} and $S^{n_{II}}$, where S^{n_I} is the $(n_I - 1)$-dimensional simplex. We can interpret x and y as vectors identifying the proportions of populations I and II playing each of the pure strategies in $\mathcal{I}$ and $\mathcal{J}$.

Let $\pi_I(x, y)$ the expected payoff to a randomly chosen agent in population I given that population proportions are (x, y).[2] In more conventional terms, this is the expected payoff to player I from using mixed

---

1. The analysis in this chapter is taken from Samuelson and Zhang [202].
2. Hence $\pi_I(x, y) = \sum_{i \in \mathcal{I}} \sum_{j \in \mathcal{J}} \pi_I(i, j) x_i y_j$.

strategy x given that the opponent uses mixed strategy y. Throughout this chapter, I shall use x and y interchangeably to denote strategy profiles or vectors of population proportions. For example, it is convenient to let $\pi_I(i, y)$ denote the payoff to a player-I agent from playing pure strategy i when the opponent plays mixed strategy y, or when y describes the strategy proportions in the opposing population. Similarly, define $\pi_{II}(x, y)$ to be the expected payoff to a randomly chosen agent in population II.

The central concept in this chapter is a selection dynamic, which is simply a differential equation. I interpret this differential equation as having been derived from an underlying model of an evolutionary process, which I shall refer to as the "selection process." In chapter 3, for example, the selection process was the Aspiration and Imitation model, and the resulting selection dynamic was a version of the replicator dynamics. When thinking of the underlying selection process as involving learning, the resulting differential equation will also be described as a learning dynamic.

Definition 4.1   Let $f : S^{n_I} \times S^{n_{II}} \to \mathbb{R}^{n_I}$ and $g : S^{n_I} \times S^{n_{II}} \to \mathbb{R}^{n_{II}}$. Then the system

$$\dot{x}_i = f_i(x, y) \qquad i = 1, \ldots, n_I \tag{4.1}$$

$$\dot{y}_j = g_j(x, y) \qquad j = 1, \ldots, n_{II} \tag{4.2}$$

is a *selection dynamic* if it satisfies, for all $(x, y) \in S^{n_I} \times S^{n_{II}}$,

(i) f and g are Lipschitz-continuous;[3]
(ii) $\sum_{i=1}^{n_I} f_i(x, y) = 0 = \sum_{j=1}^{n_{II}} g_j(x, y)$;
(iii) $x_i = 0 \Rightarrow f_i(x, y) \geq 0$, $y_j = 0 \Rightarrow g_j(x, y) \geq 0$.

Condition (i) ensures that the differential equation describing the selection process has a unique solution, although a weaker, local version of this condition would also suffice. Conditions (ii)–(iii) ensure that these dynamics do not take the system out of the appropriate simplex, hence ensuring that we can interpret a state of the process as a description of strategy proportions in the two populations. Without these restrictions, the system could produce states with negative elements or vectors of elements that sum to more than one, precluding an interpretation as vectors of population proportions.

---

3. Hence $\exists k \in \mathbb{R}_+$ such that $\forall x, x' \in S_{n_I}$, $\forall y, y' \in S_{n_{II}}$, $\|(f(x, y), g(x, y)) - (f(x', y'), g(x', y'))\| \leq k \|(x, y) - (x', y')\|$.

I shall write $x(t)$ and $y(t)$ to denote the time t values of x and y, but will suppress t whenever possible. Note that f_i/x_i is the growth rate of the proportion of population I playing strategy i. It will be helpful to define f/x to denote the vector of growth rates $(f_1/x_1, \ldots, f_{n_I}/x_{n_I})$. We shall be interested in selection dynamics that satisfy the following properties:

Definition 4.2   A selection dynamic (f, g) is *regular* if, for any state (x, y) with $x_i = 0$, there is a finite value, denoted by $f_i/0$, such that

$$\frac{f_i}{0} = \lim_{n \to \infty} \frac{f_i(x^n, y^n)}{x_i^n}$$

for any sequence $\{(x^n, y^n)\}$ with $\lim_{n \to \infty}(x^n, y^n) = (x, y)$ and with $x_i^n > 0$; with a similar condition holding for player II.

A selection dynamic (f, g) is *monotonic* if, for all $i, i' \in I$,

$$\pi_I(i, y) > (=) \pi_I(i', y) \implies \frac{f_i(x, y)}{x_i} > (=) \frac{f_{i'}(x, y)}{x_{i'}}, \tag{4.3}$$

with a similar condition for player II.

A selection dynamic is *aggregate monotonic* if, for all $p, p' \in S^{n_I}$,

$$\pi_I(p, y) > (=) \pi_I(p', y) \implies \sum_{i=1}^{n_I}(p_i - p'_i)\frac{f_i(x, y)}{x_i} > (=) 0, \tag{4.4}$$

with a similar condition for player II.

Regularity ensures that the growth rates f/x and g/y, which are continuous on the interior of $S^{n_I} \times S^{n_{II}}$ as a result of definition 4.1, are continuous on all of $S^{n_I} \times S^{n_{II}}$. This in turn ensures that if $x(0)$ and $y(0)$ are strictly positive, then $x(t)$ and $y(t)$ are strictly positive for all t. Hence strategies never literally become extinct, though the proportion of a population playing a strategy can approach zero. In cases where $x(0)$ and $y(0)$ are strictly positive, I shall refer to the initial conditions as "strictly interior" or "completely mixed." Notice that regularity and monotonicity together have the implication that if $x_i = 0$, then $f_i(x, y) = 0$. Hence the dynamics cannot introduce new strategies into the population even if they would give higher payoffs than all strategies currently being played. This is not a surprising property for a model based on imitation; though it precludes best-response dynamics. While imitation may be important, the assumption that people playing a game can never have a new idea, in the sense that they can never introduce a strategy that is currently not played, is severe.

Section 4.4 accordingly examines the effect of introducing mutations that continually inject all strategies into the system.

Monotonicity ensures that if pure strategy i receives a higher payoff than i' and the proportion of the population playing both strategies is positive, then x_i grows faster than $x_{i'}$. A monotonicity condition lies at the heart of this type of analysis because it provides the link between the payoffs of the game and the dynamics of the selection process.[4] A loose interpretation of monotonicity is that, on average, players are able to switch from worse to better strategies. Regularity and monotonicity together ensure that if one takes a sequence of states along which the expected payoff of strategy i is higher than that of i', with the expected profit difference bounded away from zero, then the difference in growth rates of x_i and $x_{i'}$ does not deteriorate to zero along this sequence. This in turn ensures that the proportion of the population playing strategy i' will approach zero.

Aggregate monotonicity requires that if (for example) the population II vector y is such that a mixed strategy p would receive a higher payoff against y than would p', then on average, those pure strategies to which p attaches relatively high probability have relatively high growth rates. By taking p and p' to attach all of their probability to pure strategies, we obtain the implication that aggregate monotonicity implies monotonicity. Examples are easily constructed to show that the converse fails.[5]

It is clear that the replicator dynamics are monotonic and that the converse need not hold, so that monotonicity is a generalization of

---

4. This monotonicity property is referred to as "relative monotonicity" by Nachbar [161]. Friedman [82] applies a similar requirement to the absolute rates of change (rather than growth rates) of the population proportions attached to the various pure strategies, and refers to qualifying dynamics as "order compatible." Swinkels [231] works with "myopic" adjustment processes, by which he means processes for which $\sum_{i=1}^{n_I} \dot{x}_i \pi_I(i, y) > 0$, with a similar condition for population II. Hence the strategy adjustments of population I must have the effect of increasing the average payoff, provided that the strategy of population II is held fixed. It is immediate that monotonic selection dynamics are myopic but that the converse fails. Given the central role of monotonicity conditions in evolutionary models, it is clearly important to investigate the conditions under which selection processes give rise to approximating differential equations that are monotonic.
5. In Samuelson and Zhang [202], we incorrectly stated the definition of aggregate monotonicity without the "(=)" on each side of implication (4.4). In this case, aggregate monotonicity does not imply monotonicity, as can be seen by noting that every dynamic on the trivial game whose payoffs are all zero would then be aggregate monotonic, but would be monotonic only if every state were stationary. I am grateful to Klaus Ritzberger for bringing this to my attention.

the replicator dynamics. Aggregate monotonicity is also a generalization of the replicator dynamics, but one which retains more of the replicator structure. To explain this remark, we first recall that the replicator dynamics are given by

$$\frac{f_i(x, y)}{x_i} = \pi_I(i, y) - \sum_{k=1}^{n_I} x_k \pi_I(k, y),$$

with a similar specification for player II. We can then solve for

$$\sum_{i=1}^{n_I}(p_i - p_i')\frac{f_i(x, y)}{x_i} = \sum_{i=1}^{n_I}(p_i - p_i')\left(\pi_I(i, y) - \sum_{k=1}^{n_I} x_k \pi_I(k, y)\right)$$

$$= \pi_I(p, y) - \pi_I(p', y),$$

leading us to conclude that the replicator dynamics are aggregate monotonic.[6]

Next, we observe that aggregate monotonic dynamics are simply transformations of the replicator dynamics. Let (f^*, g^*) denote the replicator dynamics. Section 4.5 contains the proof of the following, taken from Samuelson and Zhang [202]:

Proposition 4.1 If (f, g) is a regular, aggregate monotonic selection dynamic, then there exist functions $\alpha(x, y): S^{n_I} \times S^{n_{II}} \to \mathbb{R}$ and $\beta(x, y): S^{n_I} \times S^{n_{II}} \to \mathbb{R}$ with $\alpha(x, y) > 0$ and $\beta(x, y) > 0$ such that

$$f(x, y) = \alpha(x, y)f^*(x, y) \tag{4.5}$$

$$g(x, y) = \beta(x, y)g^*(x, y). \tag{4.6}$$

This proposition indicates that aggregate monotonic dynamics are positive multiples of the replicator dynamics, though the multiple may depend on the current state. The absolute growth rates that prevail at any state may then differ from the replicator, but the ratios of the growth rates of strategies from a given population must be preserved. Hence the only freedom we gain in passing from the replicator dynamics to an aggregate monotonic dynamic is to alter the rates at which the two populations learn.[7]

---

6. Ritzberger and Weibull [190] work with a generalization of aggregate monotonicity which they refer to as "sign-preserving dynamics." A dynamic is sign-preserving if the growth rate of strategy i is positive (negative) if and only if the payoff of strategy i is greater than (less than) the average payoff in the population.

7. The proposition and its proof involve only local arguments, and hence can be extended to the statement that if a selection dynamic is locally aggregate monotonic then it is locally a multiple of the replicator dynamics.

4.2 Equilibrium

What are the equilibrium implications of monotonic dynamics? It is immediate from the definitions that a Nash equilibrium must be a stationary state of a monotonic dynamic.[8] In particular, all strategies played with positive probability in a Nash equilibrium must receive the same payoff, which suffices for a state to be a stationary state of a monotonic dynamic. Chapter 2 noted that the replicator dynamics provide examples of cases in which stationary states under a monotonic dynamic are not Nash equilibria.

If a state is stationary under a monotonic dynamic but is not a Nash equilibrium, then it cannot be stable, yielding yet another instance of the result that "stability implies Nash." By straightforward intuition, the superior replies that prevent a strategy profile from being a Nash equilibrium in the underlying game introduce pressures that disrupt stability in the associated dynamic. In particular, if (x, y) is a stationary state but is not a Nash equilibrium because player I has a strategy i that is a better response than x to player II's strategy y, then there must exist a neighborhood of (x, y) with the property that i earns a strictly higher payoff in this neighborhood than does at least one strategy, say i', in the support of x. Furthermore, this neighborhood can be chosen so that the payoff difference between i and i' is bounded away from zero. This in turn implies that the growth rate of strategy i under a monotonic, regular dynamic must exceed that of i' on this neighborhood, with the difference again bounded away from zero. But if this is the case, then any trajectory that remains in the vicinity of (x, y) must cause the proportion of the population playing strategy i' to go to zero, a contradiction to the claim that (x, y) is a stationary state. Thus we have[9]

Proposition 4.2 Let (f, g) be a regular, monotonic selection dynamic and let (x^*, y^*) be stable. Then (x^*, y^*) is a Nash equilibrium.

How much more can we gain if we work with asymptotic stability rather than stability? Friedman [82] confirms that asymptotically stable

---

8. "Nash equilibrium" here has the obvious meaning of a state in which a positive proportion of one of the populations is attached to a strategy only if it is a best response to the opposing population.

9. Somanathan [224] establishes a partial converse of this theorem by establishing a sufficient condition for the stability of pure-strategy Nash equilibria under monotonic dynamics.

states must be Nash equilibria. As noted in chapter 2, Hofbauer and Sigmund [119, p. 282, exercise 1] have observed that stationary states of the replicator dynamics cannot be mixed-strategy Nash equilibria with full support; Ritzberger and Vogelsberger [189] and Ritzberger and Weibull [190] (see also Ritzberger [188]) have shown that asymptotically stable states of the replicator dynamics must be strict Nash equilibria. We can establish the following extension of this result to monotonic dynamics, again showing that asymptotically stable states must satisfy a strictness property.

Proposition 4.3   Let (f, g) be a regular, monotonic selection dynamic and let (x^*, y^*) be asymptotically stable. Then (x^*, y^*) is a Nash equilibrium and there does not exist $x' \in S^{n_I}$ with $x' \neq x^*$ such that

$$\pi_I(x', y^*) = \pi_I(x^*, y^*)$$

$$\pi_{II}(x', j) = \pi_{II}(x', j') \qquad \forall j, j' \in \text{supp } y^*.$$

Hence there must exist no alternative best replies for player I with the property that player II is indifferent over the strategies in the support of y^*, given this alternative strategy. To get an idea of the strength of this result, notice that if we have a pure strategy Nash equilibrium that is asymptotically stable, then it must be a strict Nash equilibrium.

The proof of this statement is immediate from the observation that if such an alternative best reply x' existed, then we can create a state $(x'', y^*) = ((1 - \epsilon)x^* + \epsilon x', y^*)$ arbitrarily close to (x^*, y^*) (by taking ϵ small). We could interpret this as a state in which most agents in population I play x^* but some play x'. Because x' is an alternative best reply, all agents in population I are playing best replies, and thus we have $f(x'', y^*) = 0$. Because player II is indifferent over the strategies in the support of y^* when player I plays x^* as well as x', and hence also when I plays x'', we have $g(x'', y^*) = 0$. Hence we have a stationary point arbitrarily close to (x^*, y^*), and the latter is not asymptotically stable.[10]

An alternative perspective on strict Nash equilibria can be provided for the special case of the replicator dynamics. A state (x, y) is

---

10. This does not establish that y^* is a best response to x''. We know that y^* is a best response to x^*, but we have not precluded the possibility of strategies for player II receiving zero probability under y^* that are alternative best responses to x^* and superior responses to x''. However, regularity and monotonicity imply $y_j = 0 \Rightarrow g_j(x'', y^*) = 0$, which suffices to establish that (x^*, y^*) is stationary.

quasi-strict if it is a Nash equilibrium with the property that $x_i = 0$ implies i is not a best reply to y and $y_j = 0$ implies j is not a best response to x. Hence strategies that are not used must be inferior replies. Section 4.5 contains the proof of the following:

Proposition 4.4   Let $(x(t), y(t))$ be a trajectory produced by the replicator dynamics with a strictly interior initial condition and converging to (x^*, y^*). Suppose that

$$\int_0^\infty |x_i(t) - x_i^*| dt \leq \mathcal{X} < \infty \qquad i = 1, \ldots, n_I \tag{4.7}$$

$$\int_0^\infty |y_j(t) - y_j^*| dt \leq \mathcal{Y} < \infty \qquad j = 1, \ldots, n_{II}. \tag{4.8}$$

Then (x^*, y^*) is quasi-strict.

This proposition indicates that equilibria that are not quasi-strict can be approached by the replicator dynamics only if *every* converging evolutionary path under the replicator dynamics converges slowly. In particular, we can interpret the integrals in (4.7)–(4.8) as the cumulative error in taking the limit of an evolutionary process as an estimate of the path. Proposition 4.4 indicates that this error will be arbitrarily large if the limit is not quasi-strict.

An alternative interpretation of this result is to let $L(\|x - x^*\|, \|y - y^*\|)$ be a bounded, continuous function with $L(0, 0) = 0$, identifying the cost of an incorrect prediction of the outcome of the game, where the actual outcome is given by the path $(x(t), y(t))$ and the limit (x^*, y^*) is predicted. Then (4.7)–(4.8) are sufficient to ensure that the discounted loss given by

$$\int_0^\infty e^{-\delta t} L(\|x(t) - x^*\|, \|y(t) - y^*\|) dt$$

does not increase without bound as the discount rate δ approaches zero. If (x^*, y^*) is not quasi-strict and hence (4.7)–(4.8) fail, then patient investigators can suffer arbitrarily large losses when using the limiting outcomes of evolutionary game theory to make predictions.

The implication of these results is that we again find a solid connection between stable behavior and Nash equilibria. Just as rationality-based theories identify Nash equilibrium as a necessary condition for behavior to be stable or self-enforcing, Nash equilibrium is required for stability in the class of adaptive models considered in this chapter. As was the case with evolutionarily stable strategies or the replicator dynamics, however, the class of monotonic dynamics provides little insight into refinements of Nash equilibria. In particular, monotonic dynamics readily direct attention to strict Nash equilibria, but appear to have little to say about the cases in which equilibrium refinements become an issue, namely the gap between Nash equilibria and strict Nash equilibria.

4.3 Strictly Dominated Strategies

Some insight into the interaction between monotonic dynamics and equilibrium refinements can be gained by examining the fate of dominated strategies under monotonic dynamics. First, consider strictly dominated strategies. The following notions are familiar and intuitively clear, but some care is required in their formulation because the difference between pure and mixed strategies will be important.

A strategy $x \in S^{n_I}$ is *strictly dominated* by $x' \in S^{n_I}$ if $\pi_I(x, y) < \pi_I(x', y)$ for all $y \in S^{n_{II}}$. Let $D(X_I, X_{II})$ be the set of pure strategies in $X_1 \subset \mathcal{I}$ that are *not* strictly dominated by any pure strategy in $\mathcal{I}$, given that player II chooses strategies from $X_{II} \subset \mathcal{J}$. Let $\overline{D}_I(M_I, M_{II})$ be the set of mixed strategies in $M_I \subset S^{n_I}$ that are *not* strictly dominated by any strategies in S^{n_I}, given that player II chooses from $M_{II} \subset S^{n_{II}}$. Similar definitions apply to player II.

Definition 4.3   The strategy $i \in I$ survives *pure strict iterated admissibility* if there exist sequences of sets

$$I = X_{I0}, X_{I1}, \ldots, X_{IT}$$

$$J = X_{II0}, X_{II1}, \ldots, X_{IIT}$$

with $i \in X_{IT}$ and with

$$X_{In+1} = D_I(X_{In}, X_{IIn}) \qquad n = 1, \ldots, T - 1$$

$$X_{IIn+1} = D_{II}(X_{In}, X_{IIn}) \qquad n = 1, \ldots, T - 1$$

and where the sequence terminates, in the sense that $X_{IT} = D_I(X_{IT}, X_{IIT})$ and $X_{IIT} = D_{II}(X_{IT}, X_{IIT})$. The strategy $x \in S^{n_I}$ survives *strict iterated admissibility* if we replace D by $\overline{D}$ and X by M in this definition.

This definition is unambiguous, in the sense that every sequence satisfying the conditions of the definition yields the same set of surviving strategies (Pearce [182]). Pearce also shows that the set of strategies surviving strict iterated admissibility is nonempty and coincides with the set of rationalizable strategies in two-player games.

Monotonic selection dynamics eliminate strategies that do not survive pure strict iterated admissibility while aggregate monotonic dynamics eliminate strategies that do not survive strict iterated admissibility. In particular, section 4.5 proves[11]

Proposition 4.5.1 Let (f, g) be a regular, monotonic selection dynamic and suppose $i \in I$ does not survive pure strict iterated admissibility. Then for any trajectory $(x(t), y(t))$ with $(x(0), y(0))$ completely mixed, we have

$$\lim_{t \to \infty} x_i(t) = 0.$$

Proposition 4.5.2 Let (f, g) be a regular, aggregate monotonic selection dynamic. Let the strategy $x' \in S^{n_I}$ fail strict iterated admissibility. Then for any trajectory $(x(t), y(t))$ with $(x(0), y(0))$ completely mixed, there exists a function $\epsilon(t)$ with $\lim_{t \to \infty} \epsilon(t) = 0$ such that for every t, there exists a pure strategy $i(t)$ in the support of x' such that $x_i(t) \le \epsilon(t)$.

The intuition behind propositions 4.5.1 and 4.5.2 is straightforward. If strategy i is strictly dominated by the pure strategy i', then i must always receive a strictly lower payoff, and hence must always be characterized by a strictly lower growth rate under a monotonic dynamic. This can be the case only if the proportion of the population playing i approaches zero. Hence strategies strictly dominated by other pure strategies will be eliminated. Once the original collection of strictly dominated strategies has been nearly eliminated, the same forces operate on a new collection of strategies that is now strictly dominated (by pure strategies). Continuing in this way, a monotonic dynamic performs the iterated elimination of strategies that are strictly dominated by pure strategies.

Additional complications arise if strategy i is dominated by the mixed strategy x. Strategy i will always earn a lower payoff than some other pure strategy, but the identity of this dominating pure strategy may change over time, preventing a simple repetition of the previous argument. However, it is straightforward to construct an analogous argument for the case of an aggregate monotonic selection dynamic.

Nachbar [161] used similar considerations to show that if pure strict iterated admissibility removes all but a single strategy for each player, than a monotonic adjustment process must converge to the surviving

---

11. Vives [243] and Milgrom and Roberts [155] obtain a similar result for supermodular games.

strategy profile. Hofbauer and Weibull [120] have recently shown that a condition called "convex monotonicity" is both weaker than aggregate monotonicity and suffices for the elimination of any pure strategy that fails strict iterated admissibility. Convex monotonicity is also necessary for pure strict iterated admissibility in the sense that for every selection dynamic failing convex monotonicity, there exists a game with a pure strategy that survives in spite of being strictly dominated by a (possibly mixed) strategy. In one of Björnerstedt's examples [31], a strategy strictly dominated by a mixed strategy survives in a continuous-time dynamic that is monotonic but not aggregate monotonic, showing that the assumption of aggregate monotonicity in proposition 4.5.2 cannot be replaced with monotonicity.

Dekel and Scotchmer [72] present an example in which a strategy strictly dominated by a mixed strategy survives under a discrete-time replicator dynamic. Coupled with propositions 4.5.1 and 4.5.2, this indicates the continuous and discrete formulations can give different results. Cabrales and Sobel [55] explore this difference, showing that strictly dominated strategies can survive if the time periods for the discrete dynamic are sufficiently long that the dynamic takes relatively large jumps. In particular, they show that if one formulates the continuous and discrete dynamics so that the former is the limit of the latter as the length of a time period shrinks, then strictly dominated strategies are eliminated for all sufficiently short time periods. Weibull [248] expands upon this issue. Björnerstedt et al. [32] present an alternative perspective, formulating the discrete replicator dynamics as an overlapping generations model and showing that dominated strategies will be eliminated if the degree of generation overlap is sufficiently high.

4.4 Weakly Dominated Strategies

Propositions 4.5.1 and 4.5.2 show that monotonic dynamics yield outcomes that respect at least the iterated removal of pure strategies strictly dominated by other pure strategies. Attention now turns to weak dominance.

An immediate observation concerning weakly dominated strategies is that a monotonic process cannot converge from an interior initial condition to an outcome that places all of the probability in one population on a pure strategy dominated by another pure strategy. Suppose that pure strategy i is weakly dominated by strategy i'. Suppose further that the selection dynamic is regular and monotonic and converges to

(x^*, y^*) with $x_i^* = 1$. Then for every t, strategy i' earns a higher payoff than i. Hence we have

$$\frac{\dot{x}_{i'}}{x_{i'}} > \frac{\dot{x}_i}{x_i} \qquad (4.9)$$

for all t. Then the system cannot converge to (x^*, y^*). This argument gives

Proposition 4.6 Let (f, g) be a regular, monotonic selection dynamic and let (x^*, y^*) be a limiting outcome of (f, g) with $(x(0), y(0))$ completely mixed. Then (x^*, y^*) cannot attach unitary probability to a pure strategy that is weakly dominated by another pure strategy.

At the same time, the outcomes of regular, monotonic dynamics need not eliminate dominated strategies entirely. The difficulty here is that the difference between the growth rates in (4.9) may approach zero as the selection dynamics converge, relieving the pressure against the weakly dominated strategy i and allowing i to survive. This can be contrasted with the case of strictly dominated strategies, where the key step in showing that such strategies cannot survive is the observation that the difference in (4.9) can be bounded away from zero.

Van Damme [240, example 9.4.3] shows that there exist stable outcomes in symmetric games that attach positive probability to weakly dominated strategies. For a simple asymmetric example, consider the game shown in figure 4.1. The phase diagram for this game, under the replicator dynamics, is given in figure 4.2.[12] It is straightforward to calculate that under the replicator dynamics, any state in which player II plays L and player I mixes between T and B, with strictly positive probability attached to T, is stable and is the limiting outcome of some nontrivial trajectory. These are Nash equilibria in which player I attaches positive probability to a weakly dominated strategy B. The difficulty is that T dominates B, but only weakly, and the payoff difference between T and B disappears as population II becomes concentrated on L. It is then possible that population II can converge to L so rapidly

---

12. I have followed the standard practice of letting the row player in a bimatrix game be player I and using x to denote this population's strategy proportions. When drawing the phase diagram, it would be natural to represent player I's strategies on the horizontal axis. However, I find it helpful to orient the phase diagram so that the corner corresponding to the pure strategy combination (T, L) is the upper left corner, as it is in the bimatrix. As a result, the phase diagram measures the probability that player I plays T along the vertical axis and the probability that player II plays R along the horizontal axis. A similar convention is adopted for subsequent phase diagrams.

	L	R
T	1, 1	1, 0
B	1, 1	0, 0

Figure 4.1
Game in which stable outcomes attach probability to dominated strategies

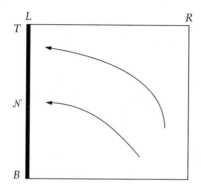

Figure 4.2
Phase diagram for figure 4.1

that the pressure pushing population *I* toward *T* dissipates too quickly to drive all population *I* to *T*, yielding an outcome in which population *I* is split between *T* and *B*.

This result directs attention to the notion of stability. A small perturbation will not prompt the system to move far away from a stable outcome; at the same time, the system need not return to the stable outcome, and may be quite unlikely to do so. In figure 4.2, for example, a perturbation towards *R* introduces new pressure away from *B* and toward *T*. The system will return to a point at which only *L* is played by population *II*, but at which a slightly higher proportion of population *I* plays *T*. This new point is stable, and the effects of a perturbation can thus remain permanently. No matter how rare mutations are, successive mutations can cause the system to move far away from a stable point. Being stable may then not be "stable enough."

One possible response is to insist on asymptotic stability. A system that is perturbed a small distance away from an asymptotically stable state returns to the asymptotically stable state. Perturbations then cannot have permanent effects, and successive, small, rare mutations cannot lead the system far away. However, we have seen that insisting on asymptotically stable states rejects pure-strategy equilibria that are not strict. One suspects that while none of these equilibria is asymptotically stable, some of them are "more stable" than others. We must then investigate the properties of equilibria that are stable but not asymptotically stable.

Asymptotic stability requires robustness to all perturbations. The appropriateness of such a requirement depends upon the source of the perturbations. If the model given by $\dot{x} = f(x, y)$ and $\dot{y} = g(x, y)$ is a literal description of the phenomenon to be studied, then we have no clues as to the likely nature of perturbations and no reason to suspect some types of perturbation as being more important than others. There accordingly appears to be no alternative to asking for robustness to all perturbations. This is especially likely to be the case in a normative analysis, where one might reasonably argue that all aspects of the problem should be captured in the model.

In a positive analysis, however, our models are never literal descriptions of the underlying process, but are instead approximations chosen in a process that attempts to balance realism and tractability. Chapter 3's justification for working with systems of differential equations as models of an evolutionary process involved an explicit approximation theorem. Perturbations are then likely to be the product of aspects of the underlying process that have been excluded from the model.

An analogous situation arises when working with tremble-based equilibrium concepts in conventional game theory. An equilibrium is trembling-hand-perfect if there are nearby equilibria in at least one converging sequence of perturbed games. But why is that particular sequence of interest? One way to avoid this question is to ask for robustness in the face of all trembles, leading to the notion of a strictly perfect equilibrium. Because such equilibria often fail to exist, Kohlberg and Mertens [135] have constructed a set-valued counterpart driven by the criterion of robustness to all perturbations. It is interesting to note that Kohlberg and Mertens also take the position that the model of the game is a complete, literal description of the strategic interaction. In this case we again have no reason to suspect some types of perturbation as being more important than others, and are accordingly prompted to ask for robustness to all perturbations. As with evolutionary models, this may be appropriate for a normative theory, but is not necessarily an appropriate standard for a positive theory, where various features of the strategic interaction excluded from the model may again cause some trembles to be more likely than others.

In light of this, we turn our attention to whether stable states of the selection dynamics will retain their stability properties when faced with certain types of perturbations that reflect unmodeled aspects of the underlying stochastic selection process. These perturbations will appear as random movements between states in the selection process and as deterministic adjustments to the differential equations that approximate that process in the long run. Suppose, for example, that the underlying process was the Aspiration and Imitation model and that we were studying the common replicator dynamics given by (2.15). Perturbations appear in the Aspiration and Imitation model in the form of mutations, which have been excluded in deriving (2.15). Taking them into consideration leads to a new long-run approximation given by (3.5).

These considerations suggest that we investigate dynamics of the following form:[13]

$$\dot{x}_i = (1 - \delta_I) f_i(x, y) + \delta_I (\zeta_i - x_i) \qquad i = 1, \ldots, n_I \tag{4.10}$$

$$\dot{y}_j = (1 - \delta_{II}) g_j(x, y) + \delta_{II} (\eta_j - y_j) \qquad y = 1, \ldots, n_{II}, \tag{4.11}$$

where (f, g) is a selection dynamic and $\zeta_i > 0$ and $\eta_j > 0$ satisfy

$$\sum_{i=1}^{n_I} \zeta_i = 1 = \sum_{j=1}^{n_{II}} \eta_j.$$

We can interpret this as a situation in which most of the population, that is, the $(1 - \delta_I)$ and $(1 - \delta_{II})$ proportions of populations I and II, evolve according to the original selection dynamic. The terms δ_I and δ_{II} can be interpreted as the proportions of the populations that are subject to perturbations and whose choices are described by the distributions ζ and η. In the Aspiration and Imitation model, these perturbations might arise out of mutations or out of unexpected realizations of the random learning process. In other models, an alternative interpretation could be that these are new entrants into the game, replacing existing players, who thus do not have experience with the game and choose their strategies according to the exogenously given ζ and η.

---

13. Hofbauer and Sigmund [119, chapter 9] present a similar model, with the perturbations motivated as transmission errors in the evolutionary process. Boylan [50] examines an analogous process in symmetric games. Bomze and Bürger [40] provide an alternative approach to stability based on perturbations.

The precise form of ζ and η will not be specified in this chapter. Given that the motivation for these terms arises out of aspects of the selection process not captured in the model, we would in general expect to gain insight into the nature of ζ and η by examining the underlying learning process. Because these elements of the underlying stochastic selection process originally seemed unimportant, we will be interested in the case in which δ_I and δ_{II} are small, and will be especially interested in the limiting case in which they approach zero.

The specification in (4.10)–(4.11) differs from that of the Aspiration and Imitation model in two respects. First, we now have two populations in an asymmetric game, and allow asymmetries across the two populations in the form of differing values of δ_I and δ_{II}. In addition, the perturbations ζ and η are assumed to be independent of the current state. Inspection of (3.5) shows that this is not the case for the Aspiration and Imitation model. This independence is convenient for the present purposes, and will be generalized in chapter 6. Notice also that the specification of the perturbations is given by ζ and η independently of the values of δ_I and δ_{II}. Although not important to the results, this makes the exposition more convenient.

We shall maintain the assumption that ζ and η are completely mixed, meaning that ζ_i and η_j are positive for all i and j. Hence the perturbations can switch agents to any of the available strategies, which is important when examining games such as the one in figure 4.1. The dominated strategy B survives among some members of the population only because R is eliminated by the selection process, and in assessing the stability of points in which B is played it is important that perturbations be able to introduce strategy R.

I shall refer to the dynamic in (4.10)–(4.11) as the "perturbed selection dynamic." We are interested in the stationary states of the perturbed selection dynamic. I shall refer to the limit of these stationary states, as the perturbations become arbitrarily small, as "limit stationary states." Section 4.5 confirms the following expected result:

Proposition 4.7   A limit stationary state of a regular, monotonic dynamic is a Nash equilibrium.

To see why we expect such a result to hold, notice that stationary states of unperturbed, regular, monotonic dynamics can fail to be Nash equilibria only because a superior reply might be played by a zero proportion of the population, and hence be unable to grow in spite of its payoff advantage. In the perturbed dynamics every strategy is always

played by at least some small fraction of the population. Any superior reply to the current state will then grow. A stationary state of the
perturbed dynamics will thus be a state where almost all agents are
playing strategies from some Nash equilibrium, while the perturbations distribute some agents among the remaining strategies. As we
approach a limit stationary state by taking the size of the perturbations to zero, the latter strategies disappear and we obtain a Nash
equilibrium.

I have drawn an analogy between (4.10)–(4.11) and tremble-based
equilibrium refinements. While the motivation for (4.10)–(4.11) and for
Selten's trembling hand perfection [208] might be similar, the two types
of perturbation have quite different effects in their respective models.
For example, weakly dominated strategies can survive in limit stationary states, unlike trembling hand perfection. In addition, it can be the
case that only under exceptional conditions does a limit stationary state
exist that attaches zero probability to dominated strategies.

To verify this last statement, consider the game in figure 4.1. Let
(f, g) be the replicator dynamic. I shall conserve on notation by letting
z denote the proportion of population I playing T and v the proportion
of population II playing R. Similarly, let ζ and η be the probabilities
attached to T and R by the perturbations. Then we have

$$\dot{z} = (1 - \delta_I)z(1 - z)v + \delta_I(\zeta - z) \tag{4.12}$$

$$\dot{v} = (1 - \delta_{II})(-v)(1 - v) + \delta_{II}(\eta - v). \tag{4.13}$$

A stationary state must satisfy $\dot{z} = 0 = \dot{v}$, and we can then solve (4.12)–
(4.13) for

$$\frac{1 - v}{z(1 - z)} = \frac{\delta_{II}}{\delta_I} \frac{1 - \delta_I}{1 - \delta_{II}} \frac{\eta - v}{z - \zeta}. \tag{4.14}$$

Now examine the limit as $\delta_I \to 0$ and $\delta_{II} \to 0$. Will the result be the
Nash equilibrium in undominated strategies, given by (T, L) and corresponding to $(v, z) = (0, 1)$? The left side of (4.14) approaches ∞ as
$(v, z) \to (0, 1)$, and $(0, 1)$ can then be the limit stationary state only if the
right side of (4.14) also approaches infinity, or only if $\delta_{II}/\delta_I \to \infty$. A
necessary (and sufficient) condition to eliminate dominated strategies
is then that population II be subject to perturbations that are arbitrarily large compared to those of population I.

Why must $\delta_{II}/\delta_I \to \infty$? Perturbations exert two conflicting forces
on population I. By causing some of population II to play R, perturbations ensure that T earns a higher payoff than B, introducing

	L	R
T	1, 1	1, 1
B	1, 1	0, 0

Figure 4.3
Game in which limit stationary states attach probability to dominated strategies

pressure toward T. At the same time, perturbations in population I simply switch some agents from T to B, introducing pressure away from T. The key is then to notice that the second force is direct and does not diminish in intensity as the system approaches the outcome $(0, 1)$. In contrast, switching agents to R introduces a pressure toward T only to the extent that it drives a wedge between the payoff of T and the average population-I payoff; inspection of (4.12) shows that the resulting pressure toward T gets weaker as more population-I agents play T. The only way to maintain enough pressure against B to eliminate the latter is then for $\delta_{II}/\delta_I \to \infty$.

This finding leads to two observations. First, there is no particular reason to believe that population II should experience more severe perturbations than population I, and certainly no reason to believe that the former perturbations will be arbitrarily more severe. Even more troubling, however, is that this result is tied closely to the game in figure 4.1. Other games will require quite different specifications of perturbations if dominated strategies are to be eliminated. For example, consider figure 4.3 and the accompanying phase diagram shown in figure 4.4. There is again a unique Nash equilibrium in undominated strategies, given by (T, L), with any state where either all of population I plays T or all of population II plays L being a Nash equilibrium.

In this case, (4.10)–(4.11) become

$$\dot{z} = (1 - \delta_I)z(1 - z)v + \delta_I(\zeta - z)$$

$$\dot{v} = (1 - \delta_{II})(-v)(1 - v)(1 - z) + \delta_{II}(\eta - v).$$

A stationary point must then satisfy

$$\frac{1 - v}{z}\frac{z - \zeta}{\eta - v} = \frac{\delta_{II}}{\delta_I}\frac{1 - \delta_I}{1 - \delta_{II}}.$$

For the perfect equilibrium (T, L), corresponding to $(v, z) = (0, 1)$, to be a limit stationary state, it must then be that in the limit as δ_I and δ_{II} approach zero,

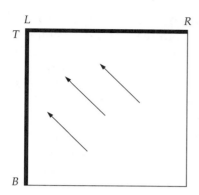

Figure 4.4
Phase diagram for figure 4.3

$$\frac{\delta_{II}}{\delta_I} \rightarrow \frac{1 - \zeta}{\eta}.$$

Any other limiting fraction yields a limit stationary outcome in which dominated strategies are played.

Perturbations will thus eliminate dominated strategies only if these perturbations are both quite special and vary in a quite specific way from game to game. A theory in which dominated strategies are eliminated must explain why the perturbations in figures 4.1 and 4.3 take the particular specifications required in those games. Considering further examples will only complicate this task. Thus the model studied in this chapter provides little evolutionary basis for eliminating dominated strategies; subsequent chapters explore the extent to which this conclusion holds in other models.

It is now clear how these perturbations differ from Selten's trembles and why they fail to eliminate dominated strategies. The perturbations in this model affect agents at their learning stage. The same perturbation process that introduces R into the population, creating pressure against the dominated strategy B, also introduces B. It is then no surprise that B is difficult to eliminate.

Samuelson and Zhang [202] examine a model in which agents are subject to perturbations, analogous to those examined by Selten [208], only in the implementation of their strategies. In this model, learning in figure 4.1 unerringly induces agents in population II to choose L and induces agents in population I to choose T, as long as T earns a higher payoff than B. However, perturbations cause some agents who intend to play T or L to actually choose B or R. By ensuring that T always earns a higher payoff than B, these perturbations yield equilibria in undominated strategies for the same reasons that Selten's trembling hand perfection [208] eliminates dominated strategies. The perturbations on the part of population II ensure that agents in population I invariably switch to an intention to play the high-payoff strategy T. In the limit as perturbations become small, actual strategies match intended strategies, giving the perfect equilibrium. This result indicates that when constructing evolutionary models, we must take care not only in specifying the technical details of perturbations (such as ζ and η in figures 4.1 and 4.3) but also in choosing how trembles affect the players.

4.5 Appendix: Proofs

Proof of Proposition 4.1   Let (f, g) be regular and aggregate monotonic. Let A and B be $n_I \times n_{II}$ matrices of player-I and player-II payoffs, where a_{ij} is the payoff to player I if player I plays his ith strategy and player II plays her jth strategy, and where b_{ij} is analogous for player II. Then

$$\pi_I(i, y) = e_i^T A y$$

$$\pi_I(x, y) = x^T A y$$

$$\pi_{II}(x, j) = x^T B e_j$$

$$\pi_{II}(x, y) = x^T B y,$$

where e_i is a vector of zeros except for a "1" in its ith place, and T denotes transposition. Given $(x, y) \in S^{n_I} \times S^{n_{II}}$, let $\zeta = (e_1^T A y, \ldots, e_{n_I}^T A y)$.

Case A   Suppose $\zeta = (c, c, \ldots, c)$ for some real number c. Then

$$\pi_I(i, y) = \pi_I(i', y) \qquad \forall i, i' \in I.$$

Then by aggregate monotonicity, we have

$$(e_i^T - e_{i'}^T) \left[\frac{f_1(x, y)}{x_1}, \ldots, \frac{f_{n_I}(x, y)}{x_{n_I}} \right] = 0,$$

which implies

$$\frac{f_i(x, y)}{x_i} = \frac{f_{i'}(x, y)}{x_{i'}} \equiv h(x, y) \qquad \forall i, i' \in I.$$

Because f is a selection dynamic, we have

$$0 = \sum_i f_i(x, y) = \sum_i x_i h(x, y) = h(x, y).$$

Therefore, for any $\alpha(x, y) > 0$, we have

$$f_i(x, y) = x_i h(x, y) = 0 = \alpha(x, y) f_i^*(x, y),$$

where the final equality holds because $\zeta = (c, c, \ldots, c)$ and f^* is the replicator dynamic.

Case B Suppose there is no $c \in \mathbb{R}$ such that $\zeta = (c, c, \ldots, c)$. By aggregate monotonicity, for all $u \in S^{n_I} - S^{n_I}$,

$$\left[\sum_i u_i = 0, \sum_i \zeta_i u_i = 0\right] \Rightarrow \sum_i u_i \frac{f_i(x, y)}{x_i} = 0.$$

Let $X = \text{span}\{u | u \in S^{n_I} - S^{n_I}, \sum_i \zeta_i u_i = 0\}$. Then the orthogonal complement of X in $\mathbb{R}^{n_I}$ is span $\{(1, \ldots, 1), \zeta\} \equiv \tilde{X}$, and we have

$$\left[\frac{f_1(x, y)}{x_1}, \ldots, \frac{f_{n_I}(x, y)}{x_{n_I}}\right] \in \tilde{X}.$$

Thus there exist $\alpha(x, y)$ and $a(x, y)$ such that

$$\left[\frac{f_1(x, y)}{x_1}, \ldots, \frac{f_{n_I}(x, y)}{x_{n_I}}\right] = \alpha(x, y)\zeta + a(x, y)(1, \ldots, 1).$$

Because $\zeta \neq (c, \ldots, c)$, there exists $u \in S^{n_I} - S^{n_I}$ such that $u^T \zeta > 0$. Then aggregate monotonicity implies

$$0 < u^T \left[\frac{f_1(x, y)}{x_1}, \ldots, \frac{f_{n_I}(x, y)}{x_{n_I}}\right] = \alpha(x, y) u^T \zeta.$$

Therefore, $\alpha(x, y) > 0$. Because (f, g) is a selection dynamic, we have

$$0 = \sum_i f_i(x, y) = \sum_i x_i(\alpha(x, y)\zeta_i + a(x, y)) = \alpha(x, y) x^T A y + a(x, y).$$

Therefore,

$$a(x, y) = -\alpha(x, y) x^T A y$$

and

$$f_i(x, y) = x_i(\alpha(x, y)\zeta_i(x, y) + a(x, y))$$

$$= \alpha(x, y) x_i [e_i^T A y - x^T A y]$$

$$= \alpha(x, y) f_i^*(x, y).$$

Combining cases A and B, we then have $\alpha(x, y) > 0$ such that $f_i(x, y) = \alpha(x, y) f_i^*(x, y)$ for all $(x, y) \in S^{n_I} \times S^{n_{II}}$. Similarly, $g_j(x, y) = \beta(x, y) g_j^*(x, y)$.   □

Proof of Proposition 4.4   Suppose (x^*, y^*) is not quasi-strict. Then there exists i' such that $x_{i'}^* = 0$, and i' is also a best reply to y^*. Consider the function defined by

$$U(t) = \prod_{i=1}^{n_I} \frac{x_i^{x_i^*}(t)}{x_{i'}(t)}.$$

Then as $t \to \infty$, $U(t) \to \infty$ (because $x_{i'}(t) \to 0$ and $\lim_{t \to \infty} \prod_{i=1}^{n_I} x_i^{x_i^*}(t) > 0$). Then

$$\frac{\dot{U}(t)}{U(t)} = \sum_{i=1}^{n_I} (x_i^* - \delta_{ii'})(e_i^T A y(t) - x^T A y(t))$$

$$= x^{*T} A y(t) - e_{i'}^T A y(t)$$

$$= (x^{*T} - e_{i'}^T) A(y(t) - y^*),$$

where the last equality appears because x^* and i' are both best replies to y^* and where $\delta_{ii'}$ ($= 1$ if $i = i'$ and $= 0$ otherwise) is the Kronecker delta. Solving this differential equation gives

$$\ln U(t) - \ln U(0) = \int_0^T (x^* - e_{i'})^T A(y(t) - y^*) dt$$

$$\leq \sum_{ij} |a_{ij}||x_i^* - \delta_{ii'}| \int_0^T |y_j(t) - y_j^*| dt$$

$$\leq \sum_{ij} |a_{ij}||x_i^* - \delta_{ii'}| \mathcal{Y} < \infty,$$

contradicting the fact that $U(t) \to \infty$ as $t \to \infty$.   □

Proof of Proposition 4.5.1   Let $I_0 \subset I$ and $J_0 \subset J$ be the sets of pure strategies for players I and II that do not survive pure strict iterated admissibility but whose population proportions do not approach zero. Assume that $I_0 \cup J_0 \neq \emptyset$ and derive a contradiction. For any strategy $\ell \in I_0 \cup J_0$, let $k(\ell)$ be defined so that $\ell \in X_{Ik(\ell)} \setminus X_{Ik(\ell)+1}$ (if $\ell \in I_0$) or $\ell \in X_{IIk(\ell)} \setminus X_{IIk(\ell)+1}$ (if $\ell \in J_0$). Let ℓ_0 be the minimizer of $k(\ell)$ on $I_0 \cup J_0$ and let $k = k(\ell_0)$. Without loss of generality, we can assume $\ell_0 \in I_0$ and can rename ℓ_0 to be i_0. Then there exists $i_1 \in I$ such that

$\pi_I(i_0, j) < \pi_I(i_1, j)$ for all $j \in X_{2k}$. Because k minimizes $k(\ell)$, we have $\lim_{t \to \infty} y_j(t) = 0$ for all $j \notin X_{2k}$. Then

$$\pi_I(i_0, y(t)) - \pi_I(i_1, y(t)) = \sum_{j \in X_{IIk}} (\pi_I(i_0, j) - \pi_I(i_1, j)) y_j(t)$$

$$+ \sum_{j \notin X_{IIk}} (\pi_I(i_0, j) - \pi_I(i_1, j)) y_j(t). \qquad (4.15)$$

As $(t \to \infty)$, the second term in (4.15) approaches zero while the first term is negative and bounded away from zero. Therefore, there exist $\epsilon > 0$ and $T > 0$ such that

$$\pi_I(i_0, y(t)) - \pi_I(i_1, y(t)) < -\epsilon \quad \forall t > T.$$

By monotonicity and regularity, we then have, for some $\delta > 0$,

$$\frac{\dot{x}_{i_0}(t)}{x_{i_0}(t)} - \frac{\dot{x}_{i_1}(t)}{x_{i_1}(t)} < -\delta \quad \forall t > T,$$

and hence $\lim_{t \to \infty} x_{i_0}(t) = 0$, contradicting the assumption that $i_0 \in I_0$.

Proof of Proposition 4.5.2 Let p be a mixed strategy for player I (without loss of generality) that fails strict iterated admissibility and let $(x(t), y(t))$ be a trajectory with $(x(0), y(0))$ completely mixed. It suffices to show that[14]

$$\lim_{t \to \infty} \prod_{i=1}^{n_I} x_i^{p_i}(t) = 0. \qquad (4.16)$$

Assume that (4.16) fails and derive a contradiction. Let A_I be the set of mixed strategies $u \in S^{n_I}$ that fail to survive iterated strict dominance and for which $\lim_{t \to \infty} \prod_{i=1}^{n_I} (x_i(t))^{u_i} \neq 0$. Let A_{II} be similarly defined for population II. Because we assume that (4.16) fails, $A_I \cup A_{II}$ is nonempty. For $a' \in A_I \cup A_{II}$, let $k(a')$ be such that $a' \in M_{Ik(a')} \setminus M_{Ik(a')+1}$ or $M_{IIk(a')} \setminus M_{IIk(a')+1}$, depending upon whether $a' \in S^{n_I}$ or $a' \in S^{n_{II}}$. Let a be the minimizer of k on $A_I \cup A_{II}$. Without loss of generality, assume $a \in A_I$ and let $k(a)$ be written simply k. Then since $a \in M_{Ik} \setminus M_{Ik+1}$, there exists $b \in S^{n_I}$ such that b strictly dominates a for all y in M_{IIk}, that is,

$$\pi_I(a, y) - \pi_I(b, y) < 0 \qquad \forall y \in M_{IIk}.$$

---

14. I use the assumption of a strictly interior initial condition here to ensure that the product in (4.16) is nonzero for any finite t.

Let Y consist of all those $y \in S^{n_{II}}$ with the property that $y_j > 0$ only if $j \in M_{IIk}$. Then

$$\pi_I(a, y) - \pi_I(b, y) < 0 \qquad \forall y \in Y.$$

By aggregate monotonicity, we have

$$\sum_{i \in S^{n_I}} (a_i - b_i) \frac{f_i(x, y)}{x_i} < 0 \qquad \forall y \in Y, x \in S^{n_I}.$$

Because Y is a closed subset of $S^{n_{II}}$, regularity ensures that there exists $\epsilon > 0$ such that

$$\sum_{i \in S^{n_I}} (a_i - b_i) \frac{f_i(x, y)}{x_i} < -\epsilon < 0 \qquad \forall y \in Y, x \in S^{n_I}. \tag{4.17}$$

Now given $y(t)$, define $\hat{y}(t)$ by

$$\hat{y}(t) = \begin{cases} \iota_t y_j(t) & \text{if } j \in M_{IIk} \\ 0 & \text{otherwise,} \end{cases}$$

where ι_t is chosen such that $\hat{y}(t) \in S^{n_{II}}$. Because $y_j(t) \to 0$ for $j \notin M_{IIk}$ (by definition of k), we have $\hat{y}(t) - y(t) \to 0$. Let

$$Z(t) = \frac{\prod_i x_i^{a_i}(t)}{\prod_i x_i^{b_i}(t)}.$$

Then differentiating gives

$$\frac{\dot{Z}(t)}{Z(t)} = \sum_{i=1}^{n_I} (a_i - b_i) \frac{\dot{x}_i(t)}{x_i(t)}$$

$$= \sum_{i=1}^{n_I} (a_i - b_i) \frac{f_i(x(t), \hat{y}(t))}{x_i(t)}$$

$$+ \sum_{i=1}^{n_I} (a_i - b_i) \left[\frac{f_i(x(t), y(t))}{x_i(t)} - \frac{f_i(x(t), \hat{y}(t))}{x_i(t)} \right]. \tag{4.18}$$

The first part of (4.18) is bounded above by $-\epsilon$ (from (4.17)), while the second part is bounded by $\epsilon/2$ when t is large. Thus there exists $T > 0$ such that

$$\frac{\dot{Z}(t)}{Z(t)} < \frac{-\epsilon}{2}$$

for $t > T$, or equivalently,

$$Z(t) < Z(T)e^{-0.5\epsilon(t-T)} \to 0.$$

Therefore, $\sum_{i=1}^{n_I} x_i^{a_i}(t) \to 0$, contradicting the fact that $a \in A_I$. □

Proof of Proposition 4.7   Let (x^*, y^*) be a limit stationary state of
(4.10)–(4.11) but not be a Nash equilibrium. Without loss of generality, let it be the case that x^* is not a best reply to y^*. Then because
the selection dynamic is monotonic and regular, there exists a pure
strategy i and a neighborhood V of (x^*, y^*) such that for all $(x, y) \in V$,
$\pi_I(i, y^*) > \pi(x^*, y^*)$, and hence $f_i(x, y)/x_i > 0$. For (x^*, y^*) to be a limit
stationary point, we must have, from (4.10),

$$0 = (1 - \delta_I^n)x_i^n \frac{f_i(x^n, y^n)}{x_i^n} + \delta_I^n(\zeta_i - x_i^n) \tag{4.19}$$

for sequences $\{\delta_I^n\}_{n=1}^{\infty}$ and $\{(x^n, y^n)\}_{n=1}^{\infty}$ with $\lim_{n\to\infty} \delta_I^n = 0$ and
$\lim_{n\to\infty}(x^n, y^n) = (x^*, y^*)$. However, $f_i(x, y)/x_i > 0$ for $(x, y) \in V$. If
$\lim_{n\to\infty} x_i^n > 0$, then notice that the second term of (4.19) becomes arbitrarily small, and hence (4.19) is strictly positive for sufficiently small
δ_I^n, a contradiction. If $\lim_{n\to\infty} x_i^n = 0$, then both terms of (4.19) are positive for sufficiently small δ_I^n, which is again a contradiction. □

5   The Ultimatum Game

In continuing our investigation of evolutionary models based on systems of deterministic differential equations, let us move now from the abstract models of the previous chapter to the consideration of a particular game, the Ultimatum Game.[1]

The Ultimatum Game is especially simple and has an unambiguous equilibrium prediction, at least if we are willing to believe in backward induction. However, there is considerable experimental evidence showing that Ultimatum Game outcomes in the laboratory do not match the unique, subgame-perfect equilibrium of the game. As this chapter will show, an evolutionary analysis can lead to conclusions that differ from those of the traditional equilibrium refinements literature and that match the qualitative nature of experimental outcomes better than does the subgame-perfect equilibrium.

There are already many explanations for observed outcomes in the Ultimatum Game. Most of these are some variant on the claim that the model is misspecified. Generally, it is the preferences of the players that are thought to be misspecified, with the players said to be affected by fairness or relative income considerations in addition to the amount of money they receive. I do not claim that such forces are absent. They may operate alongside the evolutionary forces described here, and may have a hand in explaining why the quantitative match between the model in this chapter and observed experimental outcomes is not particularly close. It may also be the case that evolutionary forces alone explain the results and the model in this chapter, while on the right track, may be the wrong evolutionary model. Yet again, evolution may have nothing to do with the results.

---

1. This chapter is taken from Binmore, Gale, and Samuelson [21].

At this point, all of these must remain possibilities. We shall assume away all other factors to demonstrate that evolutionary forces alone can yield outcomes that significantly differ from the subgame-perfect equilibrium. Questions of how we might assess the importance of evolutionary arguments, and how we might determine whether they are relevant at all, are taken up in the next chapter.

The simplicity of the Ultimatum Game makes it an ideal context in which to discuss a key issue. Evolutionary analyses are motivated by thoughts of players learning to play games. But the Ultimatum Game is so simple, what could there be to learn? This question is sufficiently compelling for some, especially when applied to responders, as to prompt statements that responders in the Ultimatum Game cannot possibly have anything to learn. I confront this issue at the end of the chapter, arguing that there are indeed things for responders to learn.

5.1   The Ultimatum Game

Consider two players with a dollar to divide. The rules of the Ultimatum Game specify that player I begins by making an offer of $x \in [0, 1]$ to player II, who then accepts or refuses. If player II accepts, player I gets $1 - x$, and player II gets x. If player II refuses, both get nothing.

This game has a unique subgame-perfect equilibrium in which player II plans to accept any offer she might receive and player I offers player II nothing. If offers must be made in multiples of some monetary unit, such as a penny, then there are other subgame-perfect equilibria, but player II gets at most a penny in any such equilibrium.

In the first of many experiments on this and related games, Güth, Schmittberger, and Schwarze [108] found that the modal offer was $\frac{1}{2}$ and that player I had roughly half a chance of being rejected if he offered about $\frac{1}{3}$ of the sum of money available. Binmore, Shaked, and Sutton [29] reported qualitatively similar results in their replication of the Ultimatum Game experiment. There have been many related studies in the interim, surveyed by Bolton and Zwick [38], Güth and Tietz [109], Roth [194] and Thaler [233].

The discussion prompted by these experimental results has proceeded in two directions. Critics of traditional economic theory have added these results to a list that includes the finitely repeated Prisoners' Dilemma and games involving the private provision of public goods; they have interpreted the results as demonstrating that the optimizing paradigm on which conventional economic models and game-theoretic equilibrium concepts are based is fundamentally mistaken. Instead, behavior is often explained in terms of social norms. Frank [80] is particularly eloquent on this subject. In bargaining games, for example, it is popular to assert that people "just play fair."

Many game theorists, on the other hand, have responded by dismissing laboratory results as irrelevant to actual behavior. The assertion is made that poorly motivated undergraduates in the artificial and unfamiliar situations often found in experiments are not reliable indicators of how people actually behave when making important economic decisions. There appears to be little doubt that experimentalists or game theorists should be cautious about making predictions unless the game is reasonably simple, the incentives are adequate, and subjects have sufficient opportunity for gaining experience with the game.[2] More and more experiments satisfy these criteria, however, and game theorists cannot ignore well-run experiments that persistently refute their predictions. In the case of the Ultimatum Game, the relevant experiments have been replicated too often for doubts about the data to persist. The theory must be reconsidered.

At first glance, the case for subgame perfection in the Ultimatum Game seems ironclad. This is a two-player game of perfect information in which each player moves only once.[3] Player I need only believe that player II will not play a weakly dominated strategy to arrive at the subgame-perfect offer. However, chapter 4 showed that evolutionary processes typically do not eliminate weakly dominated strategies, and this observation provides the motivation for an evolutionary analysis of the Ultimatum Game.

We shall assume that the replicator dynamics provide a good approximation of the long-run behavior of the learning process. Chapter 3 explains why it might be reasonable to work with *some* system of deterministic differential equations. Although there is no particular reason

2. These circumstances, stressed by Binmore [19, p. 51], appeared in chapter 3 as conditions under which initial behavior in an evolutionary process is likely to give way to strategic considerations. Smith [222] discusses the need to offer experimental subjects large incentives. Ledyard [139] examines the effects of increasing the incentives and allowing subjects to accumulate experience in experimental treatments of the private provision of public goods.

3. In particular, the criticism that subgame perfection calls for players to regard their opponents as perfectly rational after having received evidence to the contrary (cf. Binmore [18]) has no force in the Ultimatum Game.

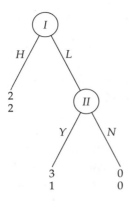

Figure 5.1
Ultimatum Minigame

	Y	N
H	2, 2	2, 2
L	3, 1	0, 0

Figure 5.2
Ultimatum Minigame, normal form

to expect the replicator dynamics to be especially applicable, it will be clear from the analysis that other regular, monotonic dynamics will give qualitatively the same results, though the specifics of the calculations will differ.

5.2   An Ultimatum Minigame

The structure of this chapter's argument is most easily isolated by examining the simplified version of the Ultimatum Game shown in figure 5.1. In this Ultimatum Minigame, player I can make a high offer (H) or a low offer (L) to player II. If he makes a high offer, it is assumed that player II accepts and the surplus of 4 is split evenly. If he makes a low offer, player II may accept (Y, for payoffs $(3, 1)$) or refuse (N, for payoffs $(0, 0)$). The normal form of this game is shown in figure 5.2.

The Ultimatum Minigame is highly stylized, but it captures the basic dilemma facing player I in the Ultimatum Game: should player I make an aggressive offer, obtaining a high payoff if it is accepted but possibly risking rejection, or should he make a less aggressive offer that yields a lower payoff when accepted but is less likely to be rejected? Alternatively, notice that rejecting an offer is a weakly dominated action in the normal form of the both the Ultimatum Game and the Ultimatum Minigame. Should player I count on player II's not taking an action that is weakly dominated and that clearly sacrifices payoff when player II has to take it?

The Ultimatum Minigame has the same strategic structure as Selten's Chain-Store Game [209], shown in figure 1.1. This section could then be interpreted as providing a possible resolution of the chain-store paradox that applies even in the case when there is just one potential entrant.

The Ultimatum Minigame has a unique subgame-perfect equilibrium S given by (L, Y) and a component $\mathcal{N}$ of Nash equilibria in which player I makes the high offer and player II chooses N with probability at least $\frac{1}{3}$. Figure 5.3 shows the trajectories of the standard replicator dynamic given by (2.15), for the case of the Ultimatum Minigame. Letting x and y be the probabilities with which N and H are played, these dynamics take the form[4]

$$\dot{y} = y(1 - y)(3x - 1) \tag{5.1}$$

$$\dot{x} = x(1 - x)(y - 1). \tag{5.2}$$

The trajectories near the component $\mathcal{N}$ are especially interesting; in particular, they resemble those near the component of Nash equilibria in figures 4.1 and 4.2. If the subgame-perfect equilibrium in the Ultimatum Minigame is to be selected, and the component $\mathcal{N}$ rejected, then the evolutionary process must lead the process away from the right end of the component $\mathcal{N}$, where the dominated strategy N is played with sufficiently high probability to make H a best response for player I, toward the left end of the component, where the dominated strategy is played relatively seldom. But we have seen in figures 4.1–4.2 that an evolutionary process need not purge dominated strategies. It is then no surprise that in the case of the Ultimatum Minigame, as shown in figure 5.3, the replicator dynamics can converge to stationary states in $\mathcal{N}$.

---

4. Once again, I have drawn this phase diagram so that the proportion of population I playing H is measured on the vertical axis and the proportion of population II playing N is measured on the horizontal axis, so that the subgame-perfect equilibrium (L, Y) occurs in the bottom left corner of both the bimatrix representation of the game and the phase diagram.

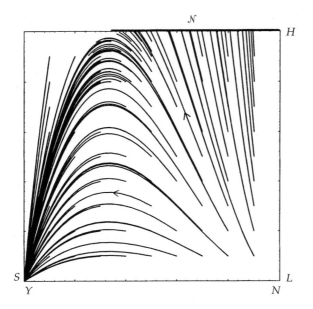

Figure 5.3
Phase diagram, no drift

We summarize the key properties of these trajectories as follows:

Proposition 5.1 The subgame-perfect equilibrium S is the unique asymptotically stable stationary state of the unperturbed replicator dynamics. With the exception of the state in which player I plays H and player II plays Y with probability $\frac{2}{3}$, the Nash equilibria in the set $\mathcal{N}$ are stable.

The replicator dynamics can thus converge to stable outcomes that are not the subgame-perfect equilibrium. However, the mere fact that interior points of $\mathcal{N}$ can attract trajectories of the replicator dynamics does not seem to be an adequate reason for regarding them as alternatives to the subgame-perfect equilibrium S. As was the case in figures 4.1–4.2, perturbations can shift the system between states in the component $\mathcal{N}$. No single, small perturbation can shift the system very far, which is to say, the states in $\mathcal{N}$ are stable. However, the effects of successive, small perturbations can accumulate to have a large, permanent impact. The survival of states in $\mathcal{N}$ is allowed only by the extinction of the strategy L. If the perturbations continually introduce the strategy L, then there is always pressure against strategy N and the fate of states in $\mathcal{N}$ is not clear.

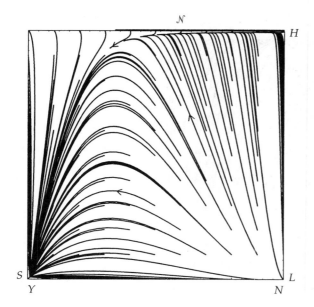

Figure 5.4
Phase diagram, comparable drift ($\delta_I = \delta_{II} = .01$)

To regard the states in $\mathcal{N}$ as serious candidates for long-run outcomes, they should then survive in the presence of such perturbations. Accordingly, we now turn to the perturbed replicator dynamics, introduced in (4.10)–(4.11) and given, for the Ultimatum Minigame, by

$$\dot{y} = (1 - \delta_I)y(1 - y)(3x - 1) + \delta_I(\tfrac{1}{2} - y) \tag{5.3}$$

$$\dot{x} = (1 - \delta_{II})x(1 - x)(y - 1) + \delta_{II}(\tfrac{1}{2} - x). \tag{5.4}$$

These equations are analogues to (5.7)–(5.8) of section 5.3 below. I call δ_I and δ_{II} the "levels of drift" in the two populations. Notice that the perturbations are assumed to introduce each of the two strategies in each population with equal frequency, reflected in the $\tfrac{1}{2}$s that appear in (5.3)–(5.4). The effect of alternative specifications will be discussed in the next section.

Figures 5.4 and 5.5 show trajectories for the perturbed replicator dynamics. These figures indicate that the outcome of the perturbed dynamics depends in an important way on the specification of the perturbations. Most of this specification has been fixed in (5.3)–(5.4), but the relative magnitudes of δ_I and δ_{II} remain open. These magnitudes

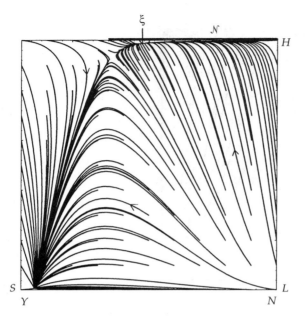

Figure 5.5
Phase diagram, more drift in population II ($\delta_I = .01$, $\delta_{II} = .1$)

determine which population, proposers or responders, experience a higher incidence of drift.

In figure 5.4, responders and proposers are equally prone to drift ($\delta_I = \delta_{II}$). In this case, there are no stationary states near the component $\mathcal{N}$. Instead, there is a single, asymptotically stable stationary state that lies very close to the subgame-perfect equilibrium and that differs from the subgame-perfect equilibrium only because the perturbations continually introduce the strategies H and N into populations I and II, and learning can accordingly never entirely eradicate such behavior. In this case of equal drift in the two populations, the perturbations inject the strategy L into population I with sufficient frequency, and hence exert enough pressure against the strategy N in population II, to ensure that stationary states near the component $\mathcal{N}$ cannot survive. Strategy N is effectively eliminated and the system converges to the subgame-perfect equilibrium.

In figure 5.5, responders are subject to more drift than proposers. In this case, there are two asymptotically stable states. Although one again lies near the subgame-perfect equilibrium, the other lies near the component $\mathcal{N}$. This state, denoted by ξ, corresponds to a Nash equilibrium in which proposers choose H. In this case, perturbations do not exert sufficient pressure against the strategy N to ensure that the subgame-perfect equilibrium appears.

To make the results suggested by these figures more precise, consider the dynamics given by (5.3)–(5.4) and let drift be small, so that (δ_I, δ_{II}) is close to $(0, 0)$. Let $\phi = \delta_{II}(1 - \delta_I)/\delta_I(1 - \delta_{II})$ be fixed and consider the limit as $(\delta_I, \delta_{II}) \to (0, 0)$ (at rates which preserve ϕ) in two cases:

Case 1: $0 < \phi < 3 + 2\sqrt{2}$.

Case 2: $3 + 2\sqrt{2} < \phi$.

Since $3 + 2\sqrt{2} \sim 5.8$, responders are subject to more drift than proposers in the second case. Section 5.6 contains the proofs of the following, including details of the specification of $\underline{x}$ and $\overline{x}$.

Lemma 5.1 Let $R(\delta_I, \delta_{II})$ be the set of stationary states of the system (5.3)–(5.4) given (δ_I, δ_{II}).

In case 1, the set $R(\delta_I, \delta_{II})$ has at most one limit point as $(\delta_I, \delta_{II}) \to (0, 0)$, which is the subgame-perfect equilibrium S.

In case 2, the set $R(\delta_I, \delta_{II})$ has three limit points as $(\delta_I, \delta_{II}) \to (0, 0)$, consisting of the subgame-perfect equilibrium S and two states contained in the component of Nash equilibria $\mathcal{N}$, in which agents from population I all play H. Agents from population II play N with probability $\underline{x}$ and $\overline{x}$ $(> \underline{x})$ in these latter two states.

Proposition 5.2 Let $A(\delta_I, \delta_{II})$ be the set of asymptotically stable states of the system (5.3)–(5.4) given values (δ_I, δ_{II}).

In case 1, the set $A(\delta_I, \delta_{II})$ has a unique limit point as $(\delta_I, \delta_{II}) \to (0, 0)$, which is the subgame-perfect equilibrium S.

In case 2, the set $A(\delta_I, \delta_{II})$ has two limit points as $(\delta_I, \delta_{II}) \to (0, 0)$, which are the subgame-perfect equilibrium S and the Nash equilibrium in which population I plays H and population II plays N with probability $\overline{x}$. (The state in which I plays H and II plays N with probability $\underline{x}$ is a limit of saddles.)

The first case gives rise to the phase diagram in figure 5.4; the second case to the phase diagram in figure 5.5. In figure 5.5, it is the asymptotically stable state ξ that converges to the Nash equilibrium in which player I plays H and player II plays Y with probability $\overline{x}$.

In the presence of drift, asymptotically stable states can thus arise near Nash equilibria of the Ultimatum Minigame that are not subgame-perfect. In the long run, we cannot restrict our attention to subgame-perfect equilibria. We have good reason to expect other Nash equilibria to appear.

How do we interpret these results? The forces behind them, including the important role played by the relative rates of drift in the two populations, are familiar from the previous chapter. In the Ultimatum Minigame, states in the component $\mathcal{N}$ are stationary because L has disappeared from the system, allowing Y and N to both be best replies, just as the strategy B persists in the game of figure 4.1 because R is eliminated. Perturbations in the Ultimatum Minigame introduce L into the population, exerting a pressure against N, but also introduce N into the population. The outcome depends upon the relative strengths of these two forces.

In figure 4.1, the perturbation introducing R is the only force exerted against strategy B. Eliminating B is then difficult, and B is entirely purged from the outcome only if the perturbations injecting R into the system are arbitrarily large compared to those injecting B. In the Ultimatum Minigame, the burden of eliminating dominated strategies does not entirely fall to perturbations. Rather than completely eliminating N, the perturbations need only reduce the frequency of N to the point that L is a best response for proposers, which is to say, perturbations need only lead to the left end of the component $\mathcal{N}$. The learning process will then lead to the subgame-perfect equilibrium. The dominated strategy N can then survive, if at all, only in sufficiently high proportion, unlike the game of figure 4.1. If the two populations are equally subject to drift, then the perturbations cannot maintain strategy N in sufficiently high frequency, instead allowing the proportion of N in the population to fall to the point that L is a best response for player I, and only the subgame-perfect equilibrium survives. The existence of asymptotically stable states near $\mathcal{N}$ then requires that responders be subject to drift at a sufficiently high rate relative to proposers.

The effects of drift, including the appearance of asymptotically stable states near the component $\mathcal{N}$, appear no matter how small are the drift parameters δ_I and δ_{II}. What is important is the relative sizes of these parameters and not their absolute value. Very small amounts of drift can have large effects on the outcome of the system because the crucial feature of drift is its behavior near the component $\mathcal{N}$. Because $\mathcal{N}$ is a component of Nash equilibria, almost all agents are playing

almost best replies when the system is near this component. The learning dynamics then exert very little pressure, and even arbitrarily small
amounts of drift can be decisive in determining whether the system is
pushed toward the component $\mathcal{N}$, yielding an asymptotically stable
state near $\mathcal{N}$, or is pushed away from $\mathcal{N}$ and toward the subgame-
perfect equilibrium.

5.3   Numerical Calculations

Let us now turn to the full Ultimatum Game. Abstract versions of the
Ultimatum Game typically allow player I to choose offers from the interval $[0, 1]$. If the game is to be actually played, as in experiments, we
must work instead with a finite set of offers. This section studies a version of the Ultimatum Game in which the players must split a "pie" of
size 40. The set of offers available to the proposer is $\mathcal{I} = \{1, 2, \ldots, 40\}$.[5]
Notice that an offer $i \in \mathcal{I}$ is the amount that player I proposes player II
should get rather than the amount player I demands for himself. An
action for player II is a choice from the set $\{Y, N\}$. Her strategies are
therefore functions $f : \{1, 2, \ldots, 40\} \to \{Y, N\}$. This is an unmanageably
large strategy set, however, and we therefore assume that player II is
restricted to functions of the form $f(i) = Y$ (for $i \geq j$) and $f(i) = N$ (for
$i < j$), for some $j \in \{1, 2, \ldots, 40\}$. We can then identify player II's strategy with the minimum acceptable offer j and the set of pure-strategy
pairs can be identified with $\mathcal{I} \times \mathcal{I}$. There are forty pure-strategy Nash
equilibria, given by (i, i), where $(i = 1, 2, \ldots, 40)$. Because $i = 0$ is excluded, the unique subgame-perfect equilibrium is $(1, 1)$.[6]

Let the fraction of proposers who make offer i at time t be denoted
by $x_i(t)$. Let $\pi_i(t)$ denote the payoff of a proposer using offer i at time
t when his opponent is drawn at random from the population of responders at time t. The average payoff in the population of proposers
at time t is then $\overline{\pi}_I(t) = x_1(t)\pi_1(t) + \cdots + x_{40}\pi_{40}(t)$. The standard replicator equation for the evolution of $x_i(t)$, given by (2.15), is

---

5. In crucial cases, solutions were also computed for games with $I = \{1, 2, \ldots, 100\}$ without significantly altering the results.

6. These strategies force player II to accept the whole pie if it is offered. They also force
player II to play strategies that are "monotonic" in the sense that player II must accept
all high offers and reject all low ones, and has available only the choice of a dividing
line between the two sets. I do not expect these restrictions to play a role in driving the
results. For an analysis of a simple case that does not impose these restrictions, see Harms
[113].

$$\dot{x}_i = x_i(\pi_i - \overline{\pi}_I) \quad i \in \{1, 2, \ldots, 40\}. \tag{5.5}$$

Similarly, let $y_j(t)$ be the fraction of responders playing strategy j at time t, with $\pi_j(t)$ being the fitness of a responder using strategy j and $\overline{\pi}_{II}(t)$ be the average fitness of responders, so that

$$\dot{y}_j = y_j(\pi_j - \overline{\pi}_{II}) \quad j \in \{1, 2, \ldots, 40\}. \tag{5.6}$$

The evolution of the system is determined by the eighty equations given by (5.5) and (5.6).

The results of chapter 4 and the analysis of the Ultimatum Minigame direct attention to the case in which (5.5)–(5.6) are subject to perturbations. It is also apparent from chapter 4 and the previous section that the manner in which these perturbations enter the model can be very important. In light of this, it is useful to have some idea of the source of the perturbation to be studied.

I view an agent as a stimulus-response mechanism with two modes of operation: a playing mode and a learning mode. An agents' playing mode operates when it is called upon to choose a strategy in one of a large number of games that it repeatedly plays against different opponents. Its behavior, when in playing mode and when confronted with a game, is triggered by a stimulus that is determined by the manner in which the game is framed. Absent the learning mode, an agent could therefore be identified with a fixed decision rule that maps a set of stimuli into a set of strategies. However, sometimes an agent will enter its learning mode between games to adjust its current decision rule. This is the equivalent of receiving the learn draw in chapter 3. When learning, the agent takes a stimulus and some information about the relative success of strategies in the game labeled by this stimulus and uses them to modify its reaction to the stimulus. The learning rule that it uses for this purpose is assumed to be fixed, and is assumed in this chapter to be approximated by the replicator dynamics.

Noise may perturb an agent either in its decision mode or in its learning mode. The perturbations in an agent's playing mode are analogous to the trembles of Selten's trembling hand perfection [208] and commonly appear in the equilibrium refinements literature. However, as chapter 1 explained, much of the noise in agents' behavior occurs in their learning process. As in chapter 4, we shall pursue the implications of noisy learning by considering only trembles in an agent's learning mode.

Let us simplify further by assuming that the only source of error lies in the possibility that an agent may mistakenly learn to play a strategy adapted to the wrong game. People make many more decisions and play many more games than they have time to analyze. These decisions must then be guided by a relatively small number of rules, and in many cases choices in a game will be guided by a rule that has been developed for use in seemingly similar but strategically different situations. In the case of a misguided proposer, we simply assume that he makes each offer i in the Ultimatum Game with some probability θ_i. If the fraction of proposers at time t who misread the game is always δ_I, and the usual arguments leading to the replicator equation apply to the fraction $1 - \delta_I$ of the proposing population who do not misread the game, then we are led to the perturbed replicator dynamics of (4.10)–(4.11), in this case given by

$$\dot{x}_i = (1 - \delta_I)x_i(\pi_i - \overline{\pi}_I) + \delta_I(\theta_i - x_i), \tag{5.7}$$

where (5.7) describes the evolution of the fraction $x_i(t)$ of agents in the proposing population who play strategy i. The corresponding equation for the population of responders is

$$\dot{y}_j = (1 - \delta_{II})x_j(\pi_j - \overline{\pi}_{II}) + \delta_{II}(\psi_j - y_j), \tag{5.8}$$

where δ_{II} is the fraction of the responding population who misread the game and ψ_j describes the choices of such agents. As with the Ultimatum Minigame, δ_I and δ_{II} are the "drift levels" of populations I and II.

What determines θ_i and ψ_j? These reflect rules of thumb or behavior learned in other games. Considerations arising in the Ultimatum Game may then tell us very little about their precise form. In this section, we shall base most of the calculations on a uniform distribution over strategies and shall look briefly at how the specification of θ_i and ψ_j affects the results.

Equations (5.7)–(5.8) have not been solved analytically, but instead by numerical calculations. Table 5.1 reports calculations for various values of δ_I and δ_{II}.[7] The rows in table 5.1 correspond to different values of δ_I; the columns correspond to different values of δ_{II}. In each

---

7. I thank John Gale for performing the calculations. The difference equation $x_i(t + \tau) - x_i(t) = \tau[(1 - \delta_I)x_i(\pi_i - \overline{\pi}) + \delta_I(\theta_i - x_i)]$ is used to approximate equation (5.7), where $\tau = 0.01$. The robustness of the approximation was tested by repeating a sample of the calculations with much smaller values of τ.

δ_{II}

	0.1	0.01	0.001	0.0001	0
0.1	7	2	1	1	1
0.01	9	7	3	1	1
0.001	9	9	7	3	1
0.0001	9	9	9	7	1
0	9	9	9	9	9

δ_I labels the rows.

Table 5.1
Fixed noise calculations

case, the system was initialized with each of the forty possible strategies being played by $\frac{1}{40}$ of each population. The mistake probabilities were also taken to be uniform, so that $\theta_i = \psi_j = \frac{1}{40}$.

The entries in table 5.1 are the modal offers made by player I after the system has converged to a point where the proportion of each population playing each strategy is unchanging in its first fifteen decimal places. In each case, the frequency with which the modal offer is played at this point is 1.00 to at least two decimal places. The equilibrium behavior of responders is much more diffuse, but is very highly concentrated on strategies less than or equal to the modal offer. Hence offers are rejected with only a very tiny probability. For example, in the cases when the modal equilibrium offer made by player I is 9, a significant fraction of responders would accept each of the offers between 1 and 9 in equilibrium (with virtually no responders insisting on more than 9)—but the fraction of responders who will refuse anything lower than 9 is high enough to make it unprofitable for proposers to reduce their offer.

Table 5.1 shows that if there is no drift ($\delta_I = \delta_{II} = 0$), then the result is an outcome in which responders receive a little more than 20% of the pie. This is a Nash equilibrium but not a subgame-perfect equilibrium of the Ultimatum Game; it is accordingly analogous to a Nash equilibrium contained in the component $\mathcal{N}$ of the Ultimatum Minigame. It is not surprising that, from some initial conditions, the system can converge to such an equilibrium. The identity of the equilibrium to which the system converges is an accident of the initial conditions. From other initial conditions, the system will converge to other Nash equilibria that are not subgame-perfect, while some initial conditions will lead to convergence to the subgame-perfect equilibrium. The uniform initial condition happens to lead to an outcome in which proposers make the offer 9.

How do these equilibria fare in the presence of drift? If the drift level among responders is sufficiently small relative to that of proposers, then the subgame-perfect equilibrium appears; if the responders' relative drift level fails to be small enough, however, then outcomes appear that are far from the subgame-perfect equilibrium. If responders are subject to enough drift compared with proposers, then player II gets a little more than 20% of the pie.[8]

This type of result is familiar from the Ultimatum Minigame, and the intuition for the result is straightforward. It is, for example, weakly dominated for player II to refuse an offer of 10%; there will therefore always be some evolutionary pressure against this strategy because, in the presence of drift, the proportion of proposers who make such low offers may be small but is continually renewed. However, if this fraction of the proposing population becomes sufficiently small, the pressure against refusals of 10% will be negligible compared with the drift engendered by the noise in the responding population. Hence, if responders experience enough drift relative to proposers, then sufficiently many responders can reject offers of 10% that it is not a best response for proposers to offer more, and we can reach outcomes that are not subgame-perfect.

Why should we anticipate that there will be at least as much drift in the population of responders as in the population of proposers—or more? Recall that drift has been motivated as a result of an agent misreading the game when learning and hence acquiring an inappropriate behavior. The context is that of a boundedly rational agent without sufficient computational power to devote full attention to all of the many games that compete for its attention. However, the frequency with which learning errors are made is unlikely to be independent of the potential costs; instead, we can expect the likelihood of a learning error to depend on how much it currently matters in payoff terms what

---

8. For the dynamic $x_i(t+1) = x_i(t) + x_i(t)(\pi_i - \bar{\pi})/\bar{\pi}$, Binmore and Samuelson [26] find results that are much the same as reported here, though they find that player I's drift level need only be at least as high as player II's in order to give subgame perfection. This difference arises because, near the subgame-perfect equilibrium, the divisor $\bar{\pi}$ becomes especially small for responders. This accentuates the learning portion of the noisy replicator dynamic, causing the responding population to be less affected by drift. The outcome for cases in which player II's drift level is higher is again 9.

strategy is played in the game. Players are likely to be more diligent in identifying games correctly when their potential gains and losses are large, and more prone to misreading games when their potential gains or losses are small. McKelvey and Palfrey [151] similarly suggest modeling players as being more likely to make mistakes or experiment with new choices if the payoff implications are small. The idea goes back to Myerson's proper equilibrium [160], which refines the idea of a trembling-hand equilibrium by making more costly mistakes less likely.[9]

In the Ultimatum Game, the result of making such an assumption is that responders will experience higher drift levels when proposers are making low offers because the responders will then have less at stake. In particular, such endogenously determined drift will lead responders to be subject to larger perturbations than proposers if the system should get close to the subgame-perfect equilibrium.

To explore this question further, we can perform calculations in which the drift levels are endogenized to bear a relationship to payoffs. Let

$$\delta_k(t) = \frac{\alpha\beta}{\alpha + \lambda_k(t)} \qquad k = I, II, \tag{5.9}$$

where α and β are constant and $\lambda_k(t)$ is the difference between the maximum and minimum of the expected payoffs attached to player k's strategies, given the current distribution of strategies in the opposing population. When this difference is zero, as is nearly the case for responders at the subgame-perfect equilibrium, the drift level takes its highest value of β. If the difference could increase all the way to infinity, $\delta_k(t)$ would decrease to zero.

The difference between the maximum and minimum payoff is an arbitrary measure of the payoffs that are at stake in a game. Calculations using alternative measures, such as the variance of the payoffs to player k's strategies, with each strategy taken to be equally likely in the variance calculation, produced analogous results (Binmore and Samuelson [26]). A more realistic measure would also incorporate a sample of past payoffs rather than employing all current payoffs, but this leads to a much more complicated process. On the other hand, I suspect that people are indeed often able to make educated guesses

---

9. Chapter 2 noted that the concept of an evolutionarily stable strategy selects proper equilibria, providing an evolutionary setting in which more costly mistakes are made less often.

α	β	Offer	$\delta_I(\infty)$	$\delta_{II}(\infty)$
10	1	9	0.26	0.52
10	0.1	9	0.024	0.053
10	0.01	9	0.0024	0.0053
10	0.001	9	0.00024	0.00053
1	1	9	0.032	0.1
1	0.1	9	0.0031	0.01
1	0.01	9	0.00031	0.001
1	0.001	9	0.000031	0.0001
0.1	1	9	0.0032	0.011
0.1	0.1	9	0.00032	0.0011
0.1	0.01	9	0.000032	0.00011
0.1	0.001	9	0.0000032	0.000011
0.01	1	9	0.00032	0.0011
0.01	0.1	9	0.000032	0.00011
0.01	0.01	9	0.0000032	0.000011
0.01	0.001	9	0.00000032	0.0000011

Table 5.2
Calculations with endogenous noise

about their compatriots' current payoffs, without necessarily being at all well informed about the strategies that secure the payoffs, and are prompted to devote energy to cases where they observe great differences in payoffs.

Table 5.2 summarizes calculations with endogenized drift for the various values of the two constants α and β listed in the first and second columns. The third column shows the modal offer made in equilibrium, meaning after the system has converged to a point where there is no additional movement in the first fifteen decimal places of the population proportions. The frequency with which the modal offer is used in equilibrium is again 1.00 to at least two decimal places, and responders' strategies range between the modal offer and zero, with virtually no rejections. The fourth and fifth columns show the drift levels in the two populations after equilibrium is achieved, which are identified as $\delta_I(\infty)$ and $\delta_{II}(\infty)$.

	1	2	3	4	5	6	7	8	9	10	11	12	13	14	
1	1	2	3	4	5	6	7	8	9	10	10	10	10	10	...
2	1	2	3	4	5	6	7	8	9	10	10	10	10	10	...
3	1	2	3	4	5	6	7	8	9	10	10	10	10	10	...
4	1	2	3	4	5	6	7	8	9	10	10	10	10	10	...
5	1	2	3	4	5	6	7	8	9	10	10	10	10	10	...
6	1	2	3	4	5	6	7	8	9	10	10	10	10	10	...
7	1	2	3	4	5	6	7	8	9	10	10	10	10	10	...
8	1	2	3	4	5	6	7	8	9	10	10	10	10	10	...
9	1	2	3	4	5	6	7	8	9	10	10	10	10	10	...
10	1	2	3	4	5	6	7	8	9	10	10	10	10	10	...
11	1	2	3	4	5	6	7	8	9	10	10	10	10	10	...
12	1	2	3	4	5	6	7	8	9	10	10	10	10	10	...
13	1	2	3	4	5	6	7	8	9	10	10	10	10	10	...
14	1	2	3	4	5	6	7	8	9	10	10	10	10	10	...
15	1	2	3	4	5	6	7	8	9	10	10	10	10	10	...

Table 5.3
Calculations with varying initial conditions

As table 5.2 shows, endogenizing drift leads to an equilibrium in which the responder population is subject to more drift than the proposer population. It is therefore not surprising that the equilibrium outcome is not subgame-perfect. The equilibrium offer is again close to 20%.

How robust are these results? First, consider the question of initial conditions. The calculations reported in tables 5.1 and 5.2 are based on a uniform initial distribution of offers over $\mathcal{I} \times \mathcal{I}$. One certainly expects the outcome of the system to be sensitive to initial conditions. Table 5.3 shows the results for some alternative initial conditions, reporting the modal equilibrium offers for the case of endogenous drift with $\alpha = 1$ and $\beta = 0.1$. The entry in row i and column j of table 5.3 is the modal equilibrium offer when the system is started from an initial condition in which all proposers play i and all responders play j. Hence the "10" that appears in cell $(15, 10)$ indicates that if the system begins will all proposers offering 15 to responders and all responders demanding at least 10, then the system converges to a state in which virtually all proposers offer 10 to responders and virtually all responders demand 10 or less. The frequency of the modal equilibrium offer remains 1.00 to at least two decimal places in each case.

Calculations were performed for each of the $40 \times 40 = 1{,}600$ possible initial conditions of this type. Table 5.3 shows some of the results, which can be summarized by the statement that the modal offer is 10 unless the responder begins the game accepting an offer less than 10, in which case the modal offer is the one the responder initially accepts. Notice that this result appears regardless of the identity of player I's initial proposal, yielding identical rows in table 5.3. The table extends downward, just as one would anticipate on the basis of its existing pattern. For cases in which proposers initially play at least 10, it extends to the right, again as one would expect—except that the cells $(38, 37)$, $(38, 38)$, $(39, 37)$, $(39, 38)$, $(39, 39)$, $(40, 37)$, $(40, 38)$, $(40, 39)$, and $(40, 40)$ yield outcomes of 9 rather than 10. For cases in which the initial offer is less than 10, the outcome is 10 as long as the initial response is not too high (generally, up to 25, though higher for lower initial proposals). Higher initial responses yield lower final outcomes. Additional calculations showed that the outcomes of initial conditions in which the proposer offers at least 10 are robust to the values of α and β.[10] The outcomes for cases in which proposers initially offered less than 10 were somewhat more sensitive, though the outcome was always at least as large as the minimum of the initial offer and initial proposal.[11]

The most striking feature of table 5.3 is the robustness of a modal equilibrium offer above 20%, which appears for a large collection of initial conditions, including all those where the initial proposal and response are at least 20%. The next section explains why we believe these are most likely to be the relevant initial conditions.

10. Using the specification of the basic dynamic as $x_i(t + 1) = x_i(t) + x_i(t)(\pi_i - \overline{\pi})/\overline{\pi}$ and taking $\lambda_k(t)$ to be the variance of the payoffs accruing to each of agent k's strategies (rather than the difference between the maximum and minimum payoff) gives similar results, though the 10s are replaced by 9s.

11. Whenever proposers initially make smaller offers than responders will accept, the dynamics begin with a race between proposers and responders, with each adjusting to match the other's strategy. The outcome of this race can be sensitive to parameters of the model when all responders initially get very low payoffs, as is the case for low proposer offers.

How do the results depend upon the specification of drift, that is, on θ_i and ψ_j? By changing these specifications, we can obtain different results. The distribution of drift among the responders is especially important. If we alter ψ_j to put relatively more weight on offers and responses near zero, equilibrium outcomes can be achieved in which the responder gets less that 20% of the pie. Causing more weight to be put on somewhat higher offers gives outcomes in which the responders get more than 20% of the pie.

It is interesting to note, however, that changing the values of θ_i and ψ_j that are attached to relatively high offers has virtually no effect on the outcome. For example, we can change the mistake probabilities so that θ_{20} and ψ_{20}, the probabilities of the "fair" offer and response, take various values up to 0.95 (with the remaining values of θ_i and ψ_j remaining equal to one another). We might view this as a case in which the rule of thumb to which most noisy players resort when not paying attention to the game is to split the pie evenly, with a minority of such players adopting completely random rules of thumb that attach equal probability to all strategies. This change leaves the results of the calculations virtually unaffected.

It is clear that we cannot place too much significance on the particular value of the equilibrium offer of a little more than 20% that repeatedly emerges in the calculations; and it is important that the results *not* be remembered for this number. Different specifications of the model can give different numbers. The important feature of the results is that the equilibrium offer is frequently far from subgame-perfect; as is the case with the Ultimatum Minigame, these results persist even when the drift levels are made very small. Reaching an outcome that is not subgame-perfect requires primarily that the responding population experience at least as much drift as the proposing population, or more. But this is the configuration of drift that appears if players are less likely to succumb to drift when making decisions that are more important.

5.4 Relevance to Experimental Data?

This chapter opened with an argument that the Ultimatum Game was interesting because of the contrast between the theoretical prediction offered by the subgame-perfect equilibrium and experimental results. How do we think the calculations of the previous section might be relevant to the experimental data? This requires a return to the distinction between the short, medium, long, and ultralong runs, restated briefly as follows:

1. In the *short run*, experimental behavior is driven primarily by norms that are triggered by the framing of the problem. If these norms have been strongly reinforced outside the laboratory, they may be hard to shift. Short-run behavior may accordingly be quite sensitive to framing effects.

2. In the *medium run*, subjects begin to learn—as emphasized by Andreoni and Miller [2], Crawford [65, 66], Miller and Andreoni [158], Roth and Erev [195, 196], and numerous others.

3. In the *long run*, this interactive learning process may converge on an equilibrium of the game.

4. In the *ultralong run*, there may be jumps between equilibria when random shocks jolt the system from one basin of attraction to another.

In the ultralong run, we can expect an evolutionary process to select the subgame-perfect equilibrium in our version of the Ultimatum Game.[12] However, given that ten periods is a typical length for experiments, while fifty periods is a relatively long experiment, it seems quite unlikely that ultralong-run considerations are relevant to experimental outcomes.

I regard the replicator dynamics as being potentially relevant to the long-run outcome of learning in the laboratory. In particular, I think the observation that learning can lead to equilibria that are not subgame-perfect, especially in a noisy environment, is an alternative to modifications of agents' preferences as an explanation of why the subgame-perfect equilibrium might not appear in experimental outcomes.

This last statement requires the consideration of two points. First, the claim that we can obtain Nash equilibria that are not subgame-perfect depends upon beginning with initial conditions or with short-run behavior that is not too close to the subgame-perfect equilibrium. Why might we expect this to be the case?

---

12. This statement relies on proposers being forced to offer responders at least 1, so that accepting is a strict best response in the subgame perfect equilibrium. The outcome is less obvious if zero offers are allowed.

The social norms triggered in the short run by laboratory experiments on the Ultimatum Game will have evolved to guide behavior in real-life bargaining situations that are superficially similar to the Ultimatum Game. We must therefore examine long-run or ultralong-run behavior in these external situations for the origin of the norms that guide short-run behavior in laboratory experiments on the Ultimatum Game.

The real-life bargaining situations shaping the norms that subjects bring to the laboratory will be complicated by informational, reputational, and other effects that are controlled away in the laboratory. The pure Ultimatum Game represents an extreme case in the class of real-life bargaining situations because all the power is on the side of the proposer. The bargaining norms that have evolved in this broader class of bargaining games, where power is more evenly shared than it is in the Ultimatum Game, will typically allocate a significant share of the surplus to the responder. We should therefore expect the initial conditions for learning in the laboratory to allocate a significant share of the surplus to the responder, and hence, as indicated in table 5.3, to lie in the basin of attraction of an outcome that offers a significant share to the responder.

Second, do laboratory experiments really provide enough experience with the game for the long run to be relevant? The appropriateness of a long-run analysis will depend greatly upon the details of the experiment. In some cases, where the framing of the problem does not invoke strongly held norms of behavior and significant incentives are provided to learn relatively easy tasks, behavior may adapt quickly and a long-run analysis may be applicable. In other cases, the behavior norms that are initially triggered may be very strongly held and will be difficult to dislodge, making the relevance of long-run results questionable. These considerations clearly place a premium on careful experimental design.

If the long run is too long, then we can ask what our calculations tell us about the medium run? Table 5.4 is a medium-run version of table 5.1. It differs from table 5.1 in that the modal offer is reported on the first occasion at which no change in consecutive iterations was detected in the first five decimal places of the fractions of proposers making each offer. (Table 5.1 does the same, but with fifteen decimal places.) The number in parentheses following each modal offer is a measure of how much learning was necessary before a temporary sta-

	δ_{II}				
	0.1	0.01	0.001	0.0001	0
0.1	9 (13)	7 (74)	7 (69)	7 (69)	7 (69)
0.01	9 (15)	9 (12)	9 (12)	9 (12)	9 (12)
0.001	9 (15)	9 (12)	9 (12)	9 (12)	9 (12)
0.0001	9 (15)	9 (12)	9 (12)	9 (12)	9 (12)
0	9 (15)	9 (12)	9 (12)	9 (12)	9 (12)

δ_I labels the rows.

Table 5.4
Medium-run calculations

bility in the first five decimal places was achieved.[13] In table 5.1, the frequency with which modal offers were used was 1.00 to at least two decimal places. The frequency with which the modal offers were used at the time reported in table 5.4 is at least .98.[14]

Table 5.4 shows that for all combinations of drift levels, and with uniform initial conditions and perturbations, the system quickly approaches a modal offer of about 20%. But table 5.1 shows this to be a medium-run result; in the long run, the system sometimes moves away to the subgame-perfect equilibrium. Only when $\delta_I < \delta_{II}$ is the medium-run behavior a useful guide to the long-run behavior of the system.

Roth and Erev [195] report Ultimatum Game simulations that spend extended periods of time, in the medium run, near equilibria that are Nash but not subgame-perfect. Explaining the experimental data on the Ultimatum Game as a set of medium-run observations of a learning process, they apply similar learning models to two other games, the Best-Shot Game and a market game, in which experimental outcomes resemble the subgame-perfect equilibrium much more closely and for which their simulations show relatively rapid convergence to the subgame-perfect equilibrium. This bolsters the hope that evolutionary models might have something to say about experimental behavior.

---

13. The measure is the number of iterations of the discrete dynamic described in note 5.3 multiplied by the step size τ, where τ of the population has an opportunity to change strategies in each iteration of the discrete equation. This measure therefore provides a crude approximation to the aggregate number of times that members of the entire population have assessed their strategies.

14. If we ask for stability in only the first three decimal places, the first line of table 5.4 would read 9 (7) 9 (6) 9 (6) 9 (6) 9 (6), with the modal offer being played with frequency at least .85 in each case. The remaining lines would be essentially unchanged.

The results of this chapter suggest that the experimental data are more than simply medium-run phenomena. Even in the long run, we need not expect to see the subgame-perfect equilibrium in the Ultimatum Game.

The conclusion I draw is that if an experiment allows the opportunity for the strategic aspects of the game to displace the norms that initially guide behavior, then we can expect Nash equilibria to appear that allocate an appreciable share of the surplus to the responder. This behavior can appear in the medium run and persist in the long run.

In the calculations presented above, the "appreciable share" is between 20% and 25% of the surplus. Responders typically receive more than this in experimental outcomes. I do not find it troubling that simply applying the replicator dynamics to the problem does not precisely duplicate experimental outcomes; we have no reason to believe that the replicator is a particularly good approximation of the way subjects learn to play the Ultimatum Game. Additional work is required to investigate what drives behavior in the Ultimatum Game. The implication of this chapter is that explanations arising out of learning in the presence of drift should take their place alongside other possibilities in accounting for this behavior.

5.5   Leaving Money on the Table

The previous sections argue that an evolutionary process based on learning can lead to a Nash equilibrium in the Ultimatum Game that is not subgame-perfect. Such an equilibrium requires that the responder be prepared to refuse low, positive offers. If offered a choice between something and nothing, such a responder would therefore sometimes choose nothing. How much learning does it take to see that something is better than nothing? What can responders be learning?

It is important to first note that the system converges to a state in which virtually all offers are accepted. Proposers learn not to make offers that are too likely to be rejected, and the fact that these offers are rarely made removes the pressure for responders to learn to accept them. Hence the strategies played by proposers are best responses. Still, sufficiently many responders must be prepared to reject more aggressive offers from proposers, should they be made, to ensure that proposers do not make such offers. When such offers are made, nothing can alter the fact that the responders are turning down money. It seemingly takes very little learning to know that this is not a good thing to do.

Because the issues appear to be so clear, behavior such as rejecting a positive offer is outlawed in conventional economic modeling. Perhaps for this reason, a common response to this chapter's argument is an incredulous "Why would anyone leave money on the table?" This comment is generally followed by an assertion that we must look to alternative models of players' preferences to explain such behavior. It is always useful to remind ourselves that all models are approximations; thus the commonly used, simple model of the Ultimatum Game is unlikely to capture players' preferences exactly. While it is important to develop richer models of preferences, it is also important to realize that responders do have something to learn in the Ultimatum Game and that the outcome of the learning process may not be a subgame-perfect equilibrium, even if players' preferences are exactly captured by the monetary payoffs of the game.

That positive offers should be refused in the short run is easy to understand. Short-run behavior in the Ultimatum Game is likely to be governed by social norms that are triggered by the framing of the laboratory experiment. Rather than being adapted to the pure Ultimatum Game, such social norms will have evolved for use in everyday relatives of the Ultimatum Game. In everyday life, however, we rarely play pure take-it-or-leave-it games. In particular, real-life games are seldom played under conditions of total anonymity; a refusal of something positive may therefore serve to maintain a reputation for toughness. For example, in the take-it-or-leave-it auction used by stores to sell their goods, a refusal of something positive may simply indicate a willingness to preserve a reputation for not taking such offers. Norms that call for refusals in commonly occurring "take it or leave it" situations therefore make good evolutionary sense. Given that such norms exist, it is unsurprising if they are sometimes inappropriately triggered in laboratory experiments. Short-run refusals of positive offers in the pure Ultimatum Game therefore create no problem for orthodox game theory.

I have argued, however, that Nash equilibria that are not subgame-perfect should be taken seriously even in the long run. This requires that responders in a noisy environment stand ready to reject *some* positive offers. On this subject, it is useful first to observe that people clearly do sometimes leave money on the table. Frank

[80], for example, reminds us that nearly everyone, including economists, leave tips in restaurants, even in restaurants they do not plan to revisit.

One may ask *why* we leave money on the table. In the case of tipping, most people are satisfied with the explanation that leaving a tip is a custom it would be uncomfortable to violate. If pressed, they might again appeal to modifications of the utility function, perhaps attributing the discomfort to the unfairness involved in disappointing the server's expectations.

Such considerations have led a number of authors to downplay strategic explanations of experimental behavior in favor of various theories of "fair play." Sophisticated versions of this approach sometimes build a taste for "fairness" into the utility functions attributed to the subjects. Ochs and Roth [181] discuss such considerations in explaining the results of an alternating offers bargaining experiment. Bolton [37] and Levine [140] explicitly construct such utility functions for this purpose.[15]

I agree that subjects find their emotions engaged in bargaining situations. They also frequently explain their bargaining behavior in the laboratory in terms of "fairness." But an approach that takes these facts at their face value is in danger of explaining too much and too little. Fairness theories explain too much because, by choosing one's fairness notion with sufficient care, one can justify a very wide range of outcomes. At the same time, such theories explain too little because they provide no insight into the origin of the fairness norms to which appeal is made.

A more fruitful approach is to ask how the custom of leaving money on the table can survive. I think the answer is quite simple. The amounts involved and the frequency with which the situation arises are too small to provide sufficient evolutionary pressure to eliminate the phenomenon in a noisy environment.

If this is the case, what of the folk explanation in terms of the discomfort felt at violating a fairness norm? In responding to such a question, it is important to appreciate that the evolutionary approach reverses the standard *explicans* and *explicandum* of the folk explanation and of an optimizing interpretation of economic theory. The players in an evolu-

---

15. Fudenberg and Levine [90] provide a method for investigating how irrational subjects must be if experimental results are generated by a process in which subjects are learning to play a self-confirming (Fudenberg and Levine [91]) equilibrium.

tionary model are not members of the species *Homo economicus*. They do not optimize relative to fixed preferences. They simply have decision rules for playing games. When a player switches from a less profitable to a more profitable strategy, he does not do so because he thinks that the switch is optimal—he is simply acting as a stimulus-response mechanism.

If people are simply stimulus-response machines, what role is left for feelings such as anger or discomfort that are often invoked to explain actions? In answering this question, we must again keep careful track of what causes what. It is easy to say that I preferred to take this foolish action rather than that wise action *because* I got angry. But angry feelings are a reflex to certain stimuli; such reflexes survive because the behaviors they induce have evolutionary advantages. Rather than seeking to explain a particular behavior in terms of the angry feelings that accompany it, we would do better to explain the angry feelings in terms of the evolutionary advantages of the behavior. In brief, being angry or fearful or amorous is how it feels to be a stimulus-response mechanism. Anger is useful because it often leads us to reject offers when there are good reasons for doing so. Unfortunately, players sometimes misread the strategic situation in which they find themselves, with the counterproductive result that an inappropriate response is triggered and reinforced by rising emotions.

Of course, none of us likes to admit that much of our behavior is little more than a set of conditioned reflexes; we prefer to offer more flattering rationalizations of the behavior. For example, the stimulus of receiving an offer of only 10% in the Ultimatum Game may be sufficiently irritating that we turn the offer down. If asked *why* we refused, we may then rationalize our behavior by arguing that irritation is an entirely appropriate response to an "unfair" offer of 10%. Such an explanation may become institutionalized and thus may reinforce the behavior that it "explains." But I see no reason to believe that fairness norms are fixed and immutable. I believe that players usually find their way to a long-run equilibrium by trial-and-error learning without having any clear understanding of the strategic realities of the game they are playing. They simply learn that certain stimulus-response behaviors are effective; after the game, they may rationalize their behavior in various ways. In bargaining experiments, they often say that the long-run equilibrium to which they found their way is fair. But, from an evolutionary perspective, how they explain their own behavior to themselves and others is beside the point. If they had found their way

to another equilibrium, they would be offering some other explanation.[16] Economists who fit a particular utility function to observed behavior would similarly find themselves proposing a different utility
function.

Because I think notions such as fairness are affected by *behavior*, I
believe that attention is most usefully focused on the evolution of behavior. If a type of behavior that prompts people to leave money on
the table survives, it will be because there is insufficient evolutionary
pressure to remove it. Fairness explanations may be offered as rationalizations of the behavior. Such stories may even be incorporated into
the workings of the stimulus-response mechanism. But the details of
how the mechanism actually works or how we explain its workings
to ourselves are secondary. The primary consideration is why some
behavior patterns survive in a population, while others necessarily
perish.

All of this leaves unanswered the question, what do agents have to
learn? I have appealed to insignificant evolutionary pressure as the reason for refusing money, but why does it take any pressure to learn to
accept something rather than nothing? What players have to learn is
that the Ultimatum Game is different from the situations in which they
usually find themselves, and that it is worthy of an independent analysis rather than the application of an "off-the-shelf" decision rule that
evolved for use in other games. I do not think the Ultimatum Game,
nor any game, is so simple that this need not be learned. Instead, people are constantly engaged in so many decisions that very few are analyzed, with most being made by applying one of a relatively short list
of existing decision rules. People have learned from experience that it
generally works better to apply a familiar, seemingly appropriate decision rule to a problem than to do a more thorough analysis.[17] Only
through experience with a game do players come to recognize it as being sufficiently different and sufficiently important to warrant its own

---

16. In Binmore et al. [30], the median long-run equilibrium claim in a laboratory implementation of the Nash Demand Game turns out to be a very good predictor of the
median claim said to be fair in a computerized postexperimental debriefing, even though
subjects are randomly chosen for an initial conditioning that directs their subsequent
play to different long-run equilibrium claims.
17. For example, most computer users have learned that when unpacking a new computer, it generally works better to plug the machine in and turn it on rather than first
read the folder prominently marked "READ ME FIRST."

analysis.[18] By the time such experience has been accumulated with the Ultimatum Game, proposers may already have learned not to make aggressive offers, and the faint incentives remaining for responders to analyze and accept such offers may be inadequate.

Do the replicator dynamics capture this learning process? Perhaps not, nor do I have an alternative model of the process. However, the replicator shares with the process the key characteristics that responders cannot be expected to automatically accept all offers, even without appealing to other models of preferences. I regard additional work on this type of learning as being very important, and I suspect that it may take us well outside the usual realm of both Bayesian and adaptive learning models.

5.6 Appendix: Proofs

Proof of Lemma 5.1 Writing $\dot{x} = \dot{y} = 0$ and $(\delta_I, \delta_{II}) = (0, 0)$ in (5.3)–(5.4) yields (x, y) equal to $(1, 0)$, $(0, 0)$ and $(x, 1)$ as candidates for the limit points of R. Recall that y is the probability with which player I plays H and x is the probability with which player II plays Y, so that $(0, 0)$ is the subgame-perfect equilibrium. The first point is a source for the unperturbed dynamics and is easily excluded as a limit point of R. We now consider the possible values of x. To this end, write $\dot{x} = \dot{y} = 0$ in (5.3)–(5.4) and then set $y = 1$. We then obtain the equation

$$\phi = \frac{x(1 - x)}{(3x - 1)(2x - 1)},$$

where $\phi = \delta_{II}(1 - \delta_I)/\delta_I(1 - \delta_{II})$. This equation has two solutions, $\underline{x}$ and $\overline{x}$, satisfying $\frac{1}{3} < \underline{x} < \sqrt{2} - 1 < \overline{x} < \frac{1}{2}$, when $\phi > 3 + 2\sqrt{2}$. In this case, $(\underline{x}, 1)$ and $(\overline{x}, 1)$ are contained in the component of Nash equilibria $\mathcal{N}$. When $\phi < 3 + \sqrt{2}$, the equation has no solutions x satisfying $0 \leq x \leq 1$. □

Proof of Proposition 5.2 The proof of the first statement is straightforward, and we consider only the second. The right side of (5.3)–(5.4) defines a function $F: \mathbb{R}^2 \to \mathbb{R}^2$ for which

---

18. Interestingly, the realization that a problem requires analysis may be more likely to appear when the game is an obvious departure from ordinary experience than with a seemingly familiar game. As a result, performance on new problems may actually be better if they are less familiar.

$DF(y, x)$

$$= \begin{bmatrix} (1 - \delta_I)(3x - 1)(1 - 2y) - \delta_I & 3(1 - \delta_I)y(1 - y) \\ (1 - \delta_{II})x(1 - x) & (1 - \delta_{II})(2x - 1)(1 - y) - \delta_{II} \end{bmatrix}.$$

The trace of this matrix is negative when $\frac{1}{3} < x < \frac{1}{2}$ and $y > \frac{1}{2}$. We therefore consider the limiting value of its determinant.

Multiply the second column of $\det DF(y, x)$ by $1 - 2x$ and then make the substitution $\delta_{II}(1 - 2x) = 2(1 - \delta_{II})x(1 - x)(1 - y)$, which holds at a rest point by virtue of (5.3)–(5.4). Factor out the term $(1 - \delta_I)(1 - \delta_{II})(1 - y)$ and write $\delta_I = \delta_{II} = 0$ and $y = 1$ in what remains. We then have to sign the determinant

$$\begin{vmatrix} 3x - 1 & 3(1 - 2x) \\ x(1 - x) & 2x^2 - 2x + 1 \end{vmatrix} = x^2 + 2x - 1.$$

The roots of the quadratic equation $x^2 + 2x - 1 = 0$ are $\sqrt{2} - 1$ and $-\sqrt{2} - 1$. It follows that the determinant is positive when $x = \bar{x} > \sqrt{2} - 1$ and negative when $x = \underline{x} < \sqrt{2} - 1$. Thus $\bar{x}$ is asymptotically stable and $\underline{x}$ is not.   □

6   Drift

6.1   Introduction

Chapter 5 argued that while simple models of learning can direct players to Nash equilibria in the Ultimatum Game in which player II receives a significant share of the surplus, this alone is insufficient to justify an interest in such equilibria. The equilibria have a compelling claim on our attention only if they satisfy an appropriate stability property. In the case of the Ultimatum Game, the stability of such equilibria initially looks doubtful. Perturbations in the offers made by proposers can only bring pressure on responders to accept offers, pushing the system toward the subgame-perfect equilibrium. We also have good reason to take such perturbations seriously. The world contains many imperfections that we conceal when formulating a game, but which reappear in the form of perturbations to any equilibrating process used to model how interactive learning leads subjects to an equilibrium of a game.

Chapter 5 then argued that in order to investigate these questions, we should expand our model to include such perturbations. Upon doing so, we find that the nature of the perturbations can be very important, and the perturbations can play a key role in *stabilizing* Nash equilibria that are not subgame-perfect.

This chapter broadens the study of perturbed equilibriating processes. Retaining the term *drift* to summarize the imperfections that may perturb an equilibriating process, the chapter investigates when drift matters and what we can say about equilibrium selection when it does.[1]

---

1. This chapter reports work in progress; for a more detailed report, including proofs, see Binmore and Samuelson [27].

The question of refining a Nash equilibrium typically arises when equilibrium strategies are accompanied by alternative best replies, which raise the question of whether the player can reasonably or "plausibly" be expected to choose her equilibrium strategy rather than some other best reply. If a refinement answers this question negatively, then it rejects the Nash equilibrium. In learning models these same alternative best replies lead to stationary states of the adjustment dynamics that occur in components rather than being isolated. The experience of the Ultimatum Minigame is typical, where we find a component containing the pure strategy equilibrium (H, N). The learning dynamics often lead toward some parts of such a component and away from other parts. As a result, the stability properties of the component hinge upon the small shocks that cause the system to drift between equilibria within the component. If this drift pushes the system toward states where the learning dynamics lead away from the component, then the component as a whole will not be stable. If the drift pushes the system toward equilibria where the learning dynamics lead toward the component, then the component will be stable, and the component of equilibrium is potentially a long-run prediction of how the game will be played.

The recognition that drift can be important might appear to vanquish all hopes for empirical applications of game theory. What good is a theory whose implications depend upon the details of perturbations that may be very small and hence very difficult to observe? To answer this question, let us return to the example set by work with nonstrategic economic theories. Conventional economics regularly works with theories involving unobservable elements; it calculates the effect of altering observable aspects of a particular model, while assuming unobservable features remain constant. Accordingly, we shall investigate the implications of manipulating the aspects of a game that we can reasonably hope to observe, namely the payoffs, while maintaining the hypothesis that the process by which players learn and the specification of drift remain constant. This leads to comparative statics predictions, concerning how the long-run outcome of play varies as the payoffs of the game vary, that may be useful in applied work. Understanding drift may then open rather than close the door to applications of game theory.

6.2   The Model

Let G denote a game with n players. As usual, we will think of this game being played by agents drawn from n populations. However, it will often be convenient to say simply that "player I plays strategy B with probability α" when the precise statement would be that "α proportion of the agents in population I play B."

Let S_ℓ be the strategy set for player ℓ. Then a state z_ℓ for population ℓ is an $|S_\ell|$-dimensional vector of nonnegative numbers whose sum equals one.[2] Such a vector is interpreted as listing the fraction of agents in population ℓ playing each of the $|S_\ell|$ pure strategies in S_ℓ. A state z of the entire system is a vector $(z_1, z_2, \ldots, z_n)$ identifying the state of each population. Let Z be the set of such states.

Selection

We now think of the choices made by the agents in this game as being determined by an evolutionary process in which agents drawn from finite populations are repeatedly matched to play the game and adjust their strategies in light of their experience. The Aspiration and Imitation model of chapter 3 is one example of such a process. This chapter will allow a broader class of models, though the assumption will be maintained that the choices resulting from this process can be described by a Markov process over the state space Z. Hence the state in which the system will next be found is a random variable that depends only upon the current strategies played by the various agents. Notice that strategy adjustments are potentially noisy, in the sense that knowing the state at time t may allow us to identify only a probability measure describing the likely identity of the next state.

We shall be focusing on the long-run behavior of our model when the populations are large. Chapter 3 indicates that the point of departure for such an analysis is a specification of the *expected* state at time $t + \tau$, given that the state at time t is $z(t) = z$. Let this expectation be given by

$$\mathcal{E}\{z(t + \tau)|z\} = F(z, \tau) + \lambda G(z, \tau). \tag{6.1}$$

To interpret (6.1), we can think of F as capturing the important forces that govern agents' strategy revisions. I refer to F as the "selection process." In a biological context, F models a process of natural

---

2. Chapter 3 generally used x to refer to a state drawn from a one-dimensional state space, while the next two chapters used x and y to refer to the strategy vectors of two populations. The analysis in this chapter is not restricted to either one dimension or two players, and uses z to denote a state.

selection driven by differences in fitness. The primary force behind the selection process here is that agents playing some strategies are likely to have more offspring than others. In the models of Young [250] and of Kandori, Mailath, and Rob [128], F models a best-response learning process driven by differences in payoffs, where the primary force driving selection is that when agents choose their strategies, they switch to best responses given the current population state. In the Aspiration and Imitation model, F models the result of agents periodically receiving payoff realizations that fall short of their aspiration level, prompting the agents to imitate others.

To study the evolutionary process, we might then begin by simply examining the selection process F, taking this to be a good model of the underlying process. For example, we could study best-response dynamics in an economic setting or the replicator dynamics in a biological setting. However, we must recognize that, like any model, the selection process is an approximation, designed to capture the important features of a problem, while excluding other considerations. Our hope when constructing the model is to include sufficiently few features of the actual situation in the specification of F as to make the model tractable, but sufficiently many to make the model a useful approximation.

The features that have been excluded from the selection mechanism F are described by G. In a biological application, G models mutations, which are random alterations in the genetic structure of an individual. In the models of Young [250] and of Kandori, Mailath, and Rob [128], G models random alterations in agents' strategies. In the Aspiration and Imitation model, G is again described as mutations, though these are viewed as mistakes in copying the strategies of others rather than biological phenomena. I shall refer to the forces captured by G as "noise."

Given that the selection mechanism F is designed to capture the important forces behind strategy choices, the events represented by G are likely to be relatively improbable. Hence we are interested in the case of small noise. The difficult issue of determining how small is "small" is generally handled by examining the limit as the events captured by the noise G become arbitrarily improbable. It is helpful to have notation that facilitates examining this limit; hence, instead of thinking of reductions in the amount of noise as changing the specification of G, we can think of G as being fixed and insert the parameter λ in (6.1). Reductions in the amount of noise are then taken to be reductions in λ. I shall refer to λ as either the "noise level" or the "drift level."[3]

The stochastic selection process and the expression in (6.1) for its expected outcome may be quite complicated. It would be helpful to have some simple techniques available for studying this process. If we assume that the derivatives of F and G exist and are bounded, then we can take a Taylor expansion of (6.1) to obtain

$$\mathcal{E}\{z(t+\tau),z\} = z + \tau[f(z) + \lambda g(z)] + O(\tau^2),\tag{6.2}$$

where f and g are the appropriate derivatives of F and G. It will be useful to bear in mind that the original stochastic process (6.1) is written as $F + \lambda G$ and not simply F. Hence the term $\lambda g(z)$ that appears in (6.2) is not the remainder from a Taylor expansion of the underlying selection process F, capturing errors that arise out of replacing F by a local approximation. Instead, we can view the selection process F itself as an approximation rather than an exact model of the underlying stochastic evolutionary process, with G and hence $\lambda g(z)$ representing factors excluded from F.

Long-Run Dynamics

To study long-run behavior, we can now rearrange (6.2) to give

$$\frac{\mathcal{E}\{z(t+\tau),z\} - z}{\tau} = f(z) + \lambda g(z) + O(\tau).\tag{6.3}$$

The next step is to follow the example of chapter 3 in taking the limits $\tau \to 0$ and $N \to \infty$, where N is the size of each of the populations. The latter limit suggests that, by the law of large numbers, we should be able to remove the expectation from the left side of (6.3); the former limit suggests we should then be able to replace the left side of (6.3) with a derivative, yielding

$$\frac{dz}{dt} = f(z) + \lambda g(z),\tag{6.4}$$

a deterministic differential equation defined on the $\prod_{\ell=1}^{n}|S_\ell|$-dimensional set given by the product of the n simplexes S_ℓ, where

---

3. This specification clearly imposes some restrictions on the way in which noise becomes small. As Bergin and Limpan [13] show, these restrictions can be important in an ultralong-run analysis. I return to this point in chapter 7. This restriction could easily be eliminated from this chapter, but only at the expense of replacing statements of the form "for all sufficiently small drift levels" with slightly more cumbersome statements.

$z(t)$ is the population state at time t. In chapter 3, the counterpart of (6.4) was the perturbed replicator dynamic given by (3.5).

The step from (6.3) to (6.4) is often justified informally. Chapter 3 showed that this step can be made formally, with the deterministic differential equation (6.4) providing an arbitrarily good long-run approximation of the behavior of the stochastic process by which strategies are adjusted, so long as we are interested in large populations.[4]

If we assume that f and g are continuously differentiable, and also Lipschitz-continuous, on an open set containing the state space Z, we ensure that there exists a unique, continuously differentiable solution $z = z(t, z(0))$, specifying the state at time t, given initial condition $z(0)$, that satisfies (6.4) (Hale [111, chapter 1]; Hirsch and Smale [118, chapter 8]). Let us further assume that the state space Z is forward-invariant under this solution, so that once the solution is an element of Z, it never leaves Z. Coupled with differentiability and the compactness of Z, this ensures that we encounter no boundary problems when working with the dynamic. Similarly, there exists a unique solution to the equation $dz/dt = f(z)$, which we can again assume renders the state space Z forward-invariant. Common examples such as the replicator dynamics satisfy these assumptions.

What are the crucial assumptions in this development? First, the underlying stochastic selection process is a Markov process, with strategy revisions depending on only the current state. This is potentially a very strong assumption, which may require players to forsake a long history of experience in order to react to an idiosyncratic current state. The extent to which this assumption is reasonable depends upon the nature of the selection process we have in mind. If agents are occasionally collecting information about others' play and choosing best responses, it is quite likely that they most readily collect information about current or recent play, and the Markov assumption may then not be too unrealistic. If strategy revisions are a response to one's own payoffs, as in the Aspiration and Imitation model, then the Markov assumption is more demanding. Much of the learning literature focuses on Markov models because of their tractability. The relaxation of this assumption is an important topic for future work.

---

4. The long-run approximation result is established in chapter 3 for the special case of a one-dimensional state space, but the argument is easily adapted to our more general case. The result is that for any time T and any ϵ, we can choose sufficiently large N and sufficiently small τ (in particular, small enough that τN^2 is small enough) that the outcome of the stochastic strategy adjustment model, at time $t \in [0, T]$, is within ϵ of the expected value given by (6.4) with probability at least $1 - \epsilon$.

The second assumption is that the differential equation (6.4) is continuously differentiable. This smoothness requirement excludes pure best-response behavior, where even an arbitrarily small difference in payoffs suffices to switch all agents to the high-payoff strategy (and in the process may make the Markov assumption more demanding). I consider this a realistic assumption. I do not think that dramatic changes in behavior are prompted by arbitrarily small differences in payoffs; instead, I expect small payoff differences to prompt small or slow shifts in behavior, with people being more likely to switch strategies as the payoff differences from doing so increase. A smooth learning process provides a good first approximation of this behavior.

If the model is well constructed, meaning that the payoffs are a good representation of preferences and f captures the important forces behind strategy revisions, then we can expect f to be closely linked to payoffs. The drift term g, however, may have little or nothing to do with the payoffs of the game. Instead, considerations excluded from the model when specifying the game and its payoffs may play a major role in shaping g. In some circumstances, drift may be completely unrelated to payoffs. This is the case in many biological models of mutation as well as in the models of Young [250] and of Kandori, Mailath, and Rob [128].

The relative importance of drift is measured by λ. I shall refer to G and λ as "noise" and the "noise level," when working with the underlying stochastic evolutionary process, but shall refer to g and λ as "drift" and the "drift level" when working with the approximation (6.4), where all of the randomness in the system has been smoothed away. In the previous two chapters, the possibility of different drift levels in different populations was emphasized; the role of λ was played by δ_I and δ_{II}. In this chapter, any such asymmetries will be captured in the specification of g.

Drift

We shall think of λ as being small, reflecting the belief that important considerations have been captured by the selection process f. This follows in the footsteps of biological models and especially the models of Young [250] and of Kandori, Mailath, and Rob [128], where the mutations that introduce noise into the models are taken to be arbitrarily improbable.

Given that we are interested in the case in which drift is very small, two questions arise. When can we simplify our analysis further by ignoring drift? This would involve replacing (6.4) with

$$\frac{dz}{dt} = f(z). \tag{6.5}$$

If drift cannot be ignored, then what can be said about the outcomes of the model?

In a biological model, ignoring drift often takes the from of ignoring the possibility of mutations and simply examining the replicator dynamics. As long as the initial conditions of the latter are modeled as being in the interior of the state space, so that all strategies are initially present, this often creates no problems. In an economic context, ignoring drift often takes the from of concentrating on the pure best-response dynamics, ignoring other forces that may perturb strategy choices. In the analysis of the Ultimatum Game of the previous chapter, ignoring drift took form of setting $\delta_I = \delta_{II} = 0$ and examining the unperturbed replicator dynamics.

The answer to the question of whether drift can be ignored depends upon what length of time we are interested in. If the ultralong run is of interest, then the question must be reformulated. To study the ultralong run, we must examine the asymptotic behavior of the original Markov process directly—rather than studying the deterministic approximation (6.4) obtained by taking expectations. The question is then not whether the drift given by g matters, but whether the noise captured by G matters. Chapter 2 noted that the stationary distribution of a Markov process can be radically changed by minute changes in one of its transition probabilities, especially if this makes a zero probability positive. Young [250] and Kandori, Mailath, and Rob [128] exploit precisely this dependence to obtain their strong, ultralong-run, equilibrium selection results.

Alternatively, we may be interested in medium-run results. To study the medium run, it suffices to fix a value of $T > 0$ and to study the behavior of (6.5) for $0 \leq t \leq T$. Over such a restricted range, the solutions to (6.4) and (6.5) are arbitrarily close if λ is sufficiently small. The following is a well-known continuity property of differential equations:

Lemma 6.1 Let $z(t, z(0), \lambda)$ solve (6.4), given initial condition $z(0)$ and drift level λ, and let $z(t, z(0), 0)$ solve (6.5) with the identical initial condition $z(0)$. Then for any $T > 0$ and $\epsilon > 0$, there exists $\lambda(T, \epsilon)$ such that for if $\lambda < \lambda(T, \epsilon)$ and $t < T$, then $\|z(t, z(0), 0) - z(t, z(0), \lambda)\| < \epsilon$.

This result indicates that we can ensure that the trajectories of the system with drift and the system without drift can be kept as close together as we want for as long as we want, provided the drift level is sufficiently small. Small amounts of drift can then be ignored.

Sánchez [203, theorem 6.3.1] provides a proof based on the following intuition. If both systems start from a common initial condition, and if λ is small, then the solutions to f and $f + \lambda g$ cannot diverge too quickly. They will diverge, however, so that f and $f + \lambda g$ will soon be evaluated at nearby but different states. As long as λ is sufficiently small, however, the Lipschitz continuity of f ensures that the behavior of the two systems is not too different when evaluated at states that are not too far apart, and hence the two systems still do not grow apart too fast. There is then an upper bound on how far apart the two solutions can grow over the time interval $[0, T]$, with this bound decreasing to zero as λ decreases to zero.

Lemma 6.1 does *not* tell us that the *long-run* behaviors of (6.4) and (6.5) are similar when λ is small. For example, the analysis of the Ultimatum Minigame of the previous chapter indicates that the asymptotic behavior of the solution to (6.5) need not approximate the asymptotic behavior of a solution to (6.4), no matter how small is λ.[5] This indicates that the limits $t \to \infty$ and $\lambda \to 0$ do not commute. First taking $\lambda \to 0$ to obtain (6.5) and then studying the limiting behavior of this system ($t \to \infty$) need not lead to the same conclusions as first studying the limiting behavior of (6.4) and then allowing $\lambda \to 0$.

The Chain-Store Game

To clarify the relationship between drift and the long run, consider the game shown in figure 6.1. Assume that the payoffs in this game satisfy the inequalities $a > e > c$, so that the player I prefers E if player II chooses A, but player I prefers N if player II chooses R. In addition, $b > d$, so that if player I chooses E, then player II prefers A. This game then has the same qualitative features as Selten's Chain-Store Game [209], shown in figure 1.1: a weakly dominated strategy for player II (given by R), a strict Nash equilibrium (given by (E, A)), and a component of weak Nash equilibria in which player I chooses

---

5. For example, (6.5) corresponds to the phase diagram shown in figure 5.3, while an appropriately specified version of (6.4) corresponds to figure 5.4. The former has a component $\mathcal{N}$ of stationary states, and potential long-run outcomes, while the latter has only S as a stationary state.

	A	R
N	e, h	e, h
E	a, b	c, d

Figure 6.1
Chain-Store Game

N and player II chooses the dominated strategy R with sufficiently high probability. If we set $a = 3$, $b = 1$, $c = d = 0$, and $e = h = 2$, then we obtain the special case of the Ultimatum Minigame, while setting $a = b = 2$, $c = d = -4$, $e = 0$, and $h = 4$ gives the Chain-Store Game.

Assume further that the dynamics $\dot{z} = f(z)$, given in (6.5), are regular and monotonic (cf. definition 4.2). The phase diagram for these dynamics then exhibits the same qualitative features as figure 5.3. In particular, the equilibrium (E, A) will be asymptotically stable; the states in the component $\mathcal{N}$ of Nash equilibria in which player I chooses N will be stable (except for the endpoint) and will attract some trajectories of the dynamics.

The analysis of the special case of the Ultimatum Minigame in the previous chapter shows that depending upon the specification of drift, the dynamics given by (6.5) can give outcomes for the game of figure 6.1 that differ significantly from those of (6.4). Figure 5.4 shows a case in which dynamic (6.4) has no stationary states near $\mathcal{N}$, unlike the unperturbed system provided by (6.5). The specification of drift lying behind figure 5.4 is shown in figure 6.2. In figure 5.5, however, the dynamic (6.4) has an asymptotically stable state ξ that is arbitrarily close to $\mathcal{N}$, reminiscent of the ability of (6.5) to converge to states in $\mathcal{N}$. The specification of drift lying behind figure 5.5 has the qualitative features shown in figure 6.3. The difference between the drift shown in figures 6.2 and 6.3 is that population II is relatively more prone to drift in figure 6.3 than in figure 6.2.[6]

Given that we are interested in the long run and not the ultralong run, why are these potentially different implications of drift relevant? Under the unperturbed dynamics of (6.5), the long-run behavior in the Chain-Store Game is that some trajectories from initial conditions converge to $\mathcal{N}$ and some do not. When the drift is as in figure 6.3, the perturbed dynamics of (6.4) give a phase diagram like the one shown

---

6. Chapter 5 indicated why this latter specification of drift may be applicable in the case of the Ultimatum Minigame.

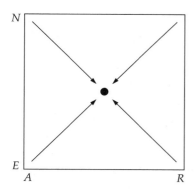

Figure 6.2
Drift for figure 5.4

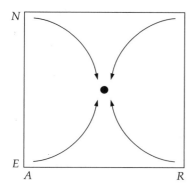

Figure 6.3
Drift for figure 5.5

in figure 5.5. Once again, the conclusion is that some trajectories converge to $\mathcal{N}$ and some do not. When drift is as in figure 6.2, the perturbed dynamics of (6.4) give a phase diagram like that of figure 5.4.
Some trajectories in figure 5.4 converge to S without coming near $\mathcal{N}$;
other trajectories do not converge to $\mathcal{N}$, but do come very close to $\mathcal{N}$.
In addition, the rate of movement of the dynamic system is very slow
near the component $\mathcal{N}$ of Nash equilibria because payoff differences
are small, and hence learning forces are weak near such a component,
while drift is also assumed to be small. As a result, the trajectories that
come close to $\mathcal{N}$ will spend a very long time near $\mathcal{N}$. Thus $\mathcal{N}$ is interesting in the medium run, even when the equilibrium that will be
selected in the long run is S. Because the medium run can be very

long, why must we go further to a long-run analysis?[7] Why do we care whether some specifications of drift lead to asymptotically stable states near $\mathcal{N}$?

The difference between the cases shown in figures 5.4 and 5.5, and the relevance of drift, hinges on what is meant by small and by long run. In figure 5.4 an argument for the relevance of $\mathcal{N}$ over a long period of time must presume that drift is sufficiently small and the initial condition places sufficient weight on the strategies H and N because otherwise the system will leave the neighborhood of $\mathcal{N}$ too quickly. The longer the time period of interest, the more stringent are these requirements. In contrast, the conditions for the applicability of $\mathcal{N}$ are much less demanding in figure 5.5, requiring only that the initial condition lie in the basin of attraction of ξ, at which point we can be confident that the system will not stray from the vicinity of $\mathcal{N}$ over any arbitrarily long time period. Studying drift is then important because drift can make an outcome such as $\mathcal{N}$ a good prediction of long-run behavior under a much wider range of circumstances.

6.3   When Can Drift Be Ignored?

The Chain-Store Game, including the special case of the Ultimatum Minigame, shows that there are cases in which drift matters. When does it matter?

Structurally Stable Stationary States

A stationary state of a differential equation is *hyperbolic* if the Jacobian matrix of the differential equation, evaluated at the stationary state, has no eigenvalues with zero real parts. Hyperbolic stationary states are isolated and are either sources, saddles, or sinks (Hirsch and Smale [118, chapter 9]). Stationary states that appear as members of connected components are thus not hyperbolic, although there are also isolated stationary states that are not hyperbolic.

For our purposes, the important characteristic of hyperbolic stationary states is that they are "structurally stable." Small perturbations of the specification of the dynamic system cannot destroy a hyperbolic stationary state and cannot alter its stability properties. In particular,

---

7. The analysis of Roth and Erev [195] is based on speed-of-adjustment arguments of this type.

proposition 6.1.1 follows immediately from the continuity of f and g on the compact set Z, while proposition 6.1.2 follows from Hirsch and Smale [118, theorems 1–2, p. 305]:[8]

Proposition 6.1.1 For any $\epsilon > 0$, there exists $\lambda(\epsilon)$ such that for any $\lambda < \lambda(\epsilon)$, every stationary state of $f + \lambda g$ lies within ϵ of a stationary state of f.

Proposition 6.1.2 Let z be a hyperbolic stationary state of f. Then $f + \lambda g$ has a hyperbolic stationary state $z(\lambda)$ that converges to z as λ converges to zero, with each $z(\lambda)$ being a sink (saddle) [source] if and only if z is a sink (saddle) [source].

Propositions 6.1.1 and 6.1.2 indicate that, if we are working with hyperbolic stationary states of f, then we can ignore drift and work with $\dot{z} = f(z)$ rather than $\dot{z} = f(z) + \lambda g(z)$. The stationary states of f provide approximate information concerning stationary states of $f + \lambda g$ that lie nearby and are of the same type. The approximation becomes arbitrarily sharp as λ gets small. The considerations that are excluded when $f + \lambda g$ is replaced by f do not have a signficant effect on the results, at least if λ is small, and might reasonably be ignored in the interest of tractability.

Nonhyperbolic stationary states do not have this structural stability. An arbitrarily small change in the dynamic system, or equivalently an arbitrarily small amount of drift, can completely change the nature of a nonhyperbolic stationary state.

For some, these observations resolve the issue of whether drift is important. It is often said that almost all dynamic systems have only hyperbolic stationary states. The Peoxito theorem (Hirsch and Smale [118, p. 314]) shows that for two-dimensional systems, there is a precise sense of "almost all" for which this statement holds; a similar result holds in higher dimensions for certain classes of dynamic systems, such as linear and gradient systems [118, pp. 313–315]. In the words of a mathematician, "Why worry about the exceptional cases" in which stationary states are not hyperbolic?

The answer is that the economics of the applications to which learning models are applied often forces us to confront nonhyperbolic stationary states. To see this, we need only recall that hyperbolic stationary states must be isolated. As a result, a Nash equilibrium that

---

8. For convenience, I shall often write simply f and $f + \lambda g$ for $\dot{z} = f(z)$ and $\dot{z} = f(z) + \lambda g(z)$, respectively.

allows alternative best replies, or that does not reach every information set in an extensive-form game (excluding some games featuring
fortuitous payoff ties), fails to be isolated under all of the familiar selection dynamics, and hence fails to be a hyperbolic stationary state. In
the Chain-Store Game, the component of equilibria $\mathcal{N}$ arises because
player I chooses not to enter the market, making the precise specification of player II's strategy immaterial (as long as it places sufficient
probability on R). Drift matters in such cases.

Unreached information sets and alternative best replies are the raw
material for equilibrium refinements. The foundation of a refinement
concept lies in the restrictions imposed on which best replies players
choose or what players do or believe at such information sets. Thus,
when equilibrium refinements are at issue, drift matters.

Asymptotically Stable Components

If we are forced by economic considerations to deal with components
of states, then perhaps a concentration on the stability of individual
stationary states is misguided. Instead, we might examine components
that satisfy some stability property. For example, consider the game
shown in figure 4.1. The single component of Nash equilibria in this
game, denoted by $\mathcal{N}$, consists of all those states in which every agent
in population II plays L. None of the stationary states corresponding
to this component is hyperbolic. All of the stationary states are stable, but none is asymptotically stable. However, as long as the initial
condition calls for some members of population II to play L, then
the regular, monotonic dynamics shown in figure 4.2 converge to $\mathcal{N}$.
In the presence of drift, the system will converge to a point close to
$\mathcal{N}$, and this point will be arbitrarily close for arbitrarily small drift
levels.

The component $\mathcal{N}$ satisfies a set version of asymptotic stability, in
that it attracts the trajectories from all nearby states. The following definition is from Bhatia and Szegö [16, definition 1.5, p. 58].

Definition 6.1 Let $z(t, z(0))$ be the solution of a differential equation defined on Z. A closed set $C \subset Z$ is *asymptotically stable* if, for
every open set $V \subset Z$ containing C, there is an open set U with $C \subset
U \subset V$ such that if $z(0) \in U$, then $z(t, z(0)) \in V$ for $t \geq 0$ and
$\lim_{t \to \infty} z(t, z(0)) \in C$.

I have mentioned that mathematicians often respond to questions
about nonhyperbolic stationary states by recommending that attention

be restricted to hyperbolic stationary states. In contrast, Guckenheimer and Holmes [107, pp. 258–259] suggest that if stationary states fail to be hyperbolic, then we should not necessarily abandon or embellish the model in an attempt to achieve hyperbolicity. Instead, we should consider the possibility that the existing model is the best description of the physical system to be studied. If so, the failure of hyperbolicity should serve not as a signal to seek a new model but as a caution to place confidence only in those features of the model which are robust to all perturbations. This last comment is analogous to the statement that if one cannot examine asymptotically stable points, then one should study asymptotically stable sets.

The asymptotic stability of $\mathcal{N}$ in the figure 4.1 is a statement about equilibrium outcomes, namely, that the payoff $(1, 1)$ will appear, rather than a statement about strategies. In particular, nothing has been said about what strategy player I will play. In other cases, different states in a component of equilibrium will correspond to differences in behavior at unreached information sets. The asymptotic stability of the component then tells us nothing about out-of-equilibrium behavior. In many cases, we will be unconcerned with such behavior because it is unobserved and has no economic consequences. In other cases, we may encounter asymptotically stable components with the property that the payoffs of at least some players vary across states in the component. We must then be concerned about which element in the component appears, for which there is no alternative to delving into the details of the drift process.

Relatively Asymptotically Stable Components

Asymptotic stability is a sufficiently strong notion that we will often be unable to find asymptotically stable components. At the same time, asymptotic stability may be stronger than necessary for a component of equilibria to be deemed worthy of attention.

Figure 6.4 shows a modification of the Chain-Store Game. The incumbent is now "tough" in the sense of Milgrom and Roberts [154] or Kreps and Wilson [138], meaning that the incumbent receives a higher payoff from resisting entry than from acquiescing. There is a unique component $\mathcal{N}$ of Nash equilibria in which the entrant does not enter and the chain store fights entry with probability at least $\frac{1}{3}$. This component is not asymptotically stable. Instead, any state in which no entry occurs is a stationary state, so that there exist stationary states arbitrarily close to $\mathcal{N}$ that are not contained in $\mathcal{N}$. However, every initial

	A	R
N	0, 4	0, 4
E	2, 2	−4, 3

Figure 6.4
Chain-Store Game, tough incumbent

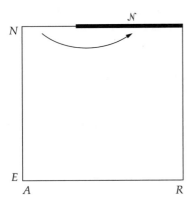

Figure 6.5
Phase diagram for Chain-Store Game with tough incumbent

condition that lies in the interior of the state space yields a trajectory that converges to $\mathcal{N}$ (and that does not stray too far away if it starts nearby). Figure 6.5 shows the phase diagram for this game.

The component $\mathcal{N}$ appears to be nearly asymptotically stable. The following definition, a slight modification of definition 5.1 of Bhatia and Szegö [16, p. 99], allows us to make this precise:

Definition 6.2 Let $z(t, z(0))$ be the solution of a differential equation on state space Z given initial condition $z(0)$. A closed set $C \subset Z$ is *asymptotically stable relative to* $W \subset Z$ if, for every open set V containing C, there is an open set U with $C \subset U \subset V$ such that if $z(0) \in U \cap W$, then $z(t, z(0)) \in V$ for $t \geq 0$ and $\lim_{t \to \infty} z(t, z(0)) \in C$.

In the case of the Chain-Store Game of figure 6.4, the set W can be taken to be the interior of the state space, in which case we say that $\mathcal{N}$ is *asymptotically stable relative to the interior*. Given the common presumption that drift points into the interior of the state space, this is an especially interesting case. We might then resolve stability difficulties

	d	a
D	2, 1	2, 1
Ad	1, 8	8, 4
Aa	1, 8	4, 32

Figure 6.6
Centipede Game

by simply seeking components that are asymptotically stable with respect to the interior.

There is reason to believe that such an approach will be quite effective in some cases. Cressman and Schlag [70] study the replicator dynamic in a class of generic extensive-form games of perfect information in which no path through the game involves more than three moves and in which there are at most two actions at each node. Such games may have multiple Nash equilibria, but Cressman and Schlag show that any such game has a unique component of equilibria that is asymptotically stable with respect to the interior, and this component corresponds to the subgame-perfect equilibrium outcome. As long as the initial condition is interior, we are then assured of convergence to the subgame-perfect equilibrium outcome.

On the other hand, Cressman and Schlag [70] present counterexamples to show that this result does not extend to longer games. More importantly, the relationship between asymptotic stability with respect to the interior and the importance of drift requires some further investigation. If a component is asymptotically stable with respect to the interior, than any perturbation of the system away from this component and into the interior of the state space will give rise to convergence back to the component. This does not imply, however, that drift is unimportant. For example, consider the Centipede Game, whose normal form is shown in figure 6.6. Figure 6.7 shows the extensive form of this game, while figure 6.8 shows the state space, indicating the direction of the dynamics in two faces of the state space spanned by $\{D, Aa\} \times \{d, a\}$ and $\{D, Ad\} \times \{d, a\}$. Notice that in the face of this state space corresponding to the strategies $\{D, Aa\} \times \{d, a\}$, the phase diagram has the same qualitative features as the Chain-Store Game. Then there exists a specification of drift such that, in this face, the phase diagram matches that of figure 5.4 and every trajectory converges to (Aa, a). There accordingly exist some specifications of drift such that some trajectories

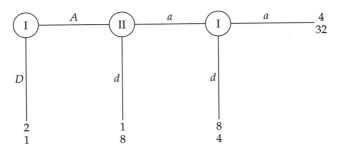

Figure 6.7
Centipede Game, extensive form space

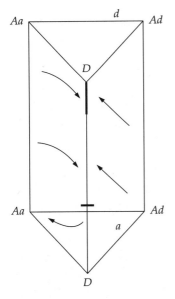

Figure 6.8
Centipede Game, state space

in the resulting dynamics with drift start arbitrarily close to the component containing the subgame-perfect equilibrium but converge to states that are far away. These trajectories, however, are confined to the face of the state space spanned by $\{D, Aa\} \times \{d, a\}$, and the drift in question never pushes the system out of this face. The same reasons prompting us to be interested in asymptotic stability with respect to the interior suggest that we should consider only trajectories with interior origins and interior-pointing drift. But lemma 6.1 can then be invoked to ensure that there are states in the interior of the state space, and arbitrarily close to the component containing the subgame-perfect equilibrium, that give rise to trajectories leading far away from that component (though they may eventually converge to the component).[9] The details of the drift process can thus be relevant, even when dealing with components that are asymptotically stable with respect to the interior.

6.4   When Drift Matters

In the Chain-Store Game of figure 6.1, a component of nonhyperbolic stationary states arises that is neither asymptotically stable nor asymptotically stable relative to the interior. In these cases, drift matters. The forces captured by $g(z)$ are not negligible for the purposes of long-run prediction, and we cannot evade studying the model $\dot{z} = f(z) + \lambda g(z)$. What can be said about such cases?

In addition to assuming that f and g are continuously differentiable, let us assume that the stationary states of $\dot{z} = f(z) + \lambda g(z)$ are hyperbolic for sufficiently small λ. This latter assumption would be natural if we were interested in a fixed value of λ. If the stationary states of $f + \lambda g$ were not hyperbolic, then we would argue that yet more unmodeled sources of drift need to be incorporated into the model. The assumption that the stationary states of $f + \lambda g$ are hyperbolic for all sufficiently small λ is stronger, but is necessary if we are to examine small drift levels without committing to how small is small. Let us simplify further by assuming that the basin of attraction of the set of all stationary points of $f + \lambda g$ is the entire space Z; this excludes cycles and other complicated or chaotic trajectories. Although cycles or even more exotic behavior must be important in some cases, it is important to begin by understanding cases in which the dynamics are relatively well behaved.

The indication that drift matters is that the unperturbed dynamic $\dot{z} = f(z)$ has stationary states that occur in a component C rather than being isolated. When does this component possess sufficient stability to be a possible prediction of long-run behavior for a learning dynamic with drift? The component can be considered a potentially relevant prediction of long-run behavior if, in the presence of drift, we can find

---

9. Ponti [183] provides numerical calculations of such trajectories.

stationary states of $\dot{z} = f(z) + \lambda g(z)$ close to C for all sufficiently small $\lambda > 0$, with these stationary states approaching C as $\lambda \to 0$ and also lying in the interior of a set that does not shrink as $\lambda \to 0$ and that is contained in the basin of attraction of the stationary states.

It will be helpful to begin with a simple example. Consider the game shown in figure 4.1. We already know that the *component* $\mathcal{N}$ of equilibria in this game is asymptotically stable. How does this component fare in the presence of drift?

It will avoid some confusion to let the state space be denoted by W, and a state by w, throughout this discussion. In addition, let the dynamics be denoted by $\tilde{f}(w)$ and $\tilde{g}(w)$, where these are assumed to be continuously differentiable, regular, and monotonic.

Let $B_\delta(\mathcal{N})$ be the set of all points in W whose distance from $\mathcal{N}$ in the max norm is δ or less.[10] Hence, writing simply B_δ for $B_\delta(\mathcal{N})$, we have $B_0 = \mathcal{N}$ and $B_1 = W$. Let $D_\delta = \partial B_\delta \cap \partial(W \setminus B_\delta)$, where ∂S denotes the boundary of a set S. Then D_δ is the set of points contained in the boundaries of both B_δ and its complement. Now suppose that for all sufficiently small $\delta > 0$, the following two conditions hold:

For all $w \in B_\delta$, $\tilde{f}(w)$ points into $B_\delta \setminus D_\delta$ at w, \hfill (6.6)

For all $w \in \mathcal{N}$, $\tilde{g}(w)$ points into B_δ^o at w, \hfill (6.7)

where S^o is the interior of a set S, and where $\tilde{f}$ is said to point into S at w if $w + \epsilon \tilde{f}(w) \in S$ for all small enough $\epsilon > 0$. Condition (6.6) imposes a restriction on the behavior of the learning dynamic $\tilde{f}$ at states in the set D_δ (and only those states), requiring that $\tilde{f}$ point into the set B_δ and that $\tilde{f}$ not point into D_δ. In the game of figure 4.1 condition (6.6) will hold for all $\delta < 1$ as long as $\tilde{f}$ is regular and monotonic. Condition (6.7) places a restriction on the drift $\tilde{g}$ on the component $\mathcal{N}$, requiring the drift point into the interior of the set B_δ. In figure 4.1 this will hold as long as g points into the interior of the state space, as is commonly assumed. Conditions (6.6)–(6.7) are illustrated, for the game of figure 4.1, in figure 6.9.

The important observation is now that g is continuous. We can then deduce from (6.7) that there exists $\delta^* > 0$ such that for all $0 < \delta \le \delta^*$, we have

$\tilde{g}(w)$ points into B_δ^o for all $w \in B_\delta \setminus D_\delta$ \hfill (6.8)

$\tilde{g}(w)$ points into W^o for all $w \in B_\delta$. \hfill (6.9)

Conditions (6.8)–(6.9) are illustrated in figure 6.10.

---

10. The max norm is defined by $\|w\| = \max_i |w_i|$.

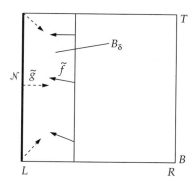

Figure 6.9
Illustration of conditions (6.6)–6.7)

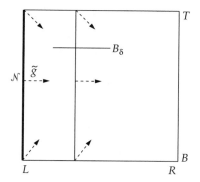

Figure 6.10
Illustration of conditions (6.8)–(6.9)

This observation allows us to conclude that if conditions (6.6)–(6.7) hold, then there exists a $\lambda(\delta)$ with the property that stationary points of $\dot{z} = \tilde{f}(w) + \lambda \tilde{g}(w)$ can be found in B_δ^o whenever $0 < \delta \leq \delta^*$ and $0 < \lambda \leq \lambda(\delta)$. To verify that such a $\lambda(\delta)$ exists, fix $\delta \leq \delta^*$. Then observe that $\tilde{f}$ is continuous on W and, for sufficiently small δ^* ($\delta^* < 1$ in the game of figure 4.1, though smaller values of δ^* may be required when we generalize to other games), has zeros in B_δ only on $\mathcal{N}$. Combined with (6.6) and (6.8), this allows us to conclude that there exists $\lambda(\delta)$ such that $\tilde{f}(w) + \lambda \tilde{g}(w)$ points into B_δ^o for all $\lambda < \lambda(\delta)$ and all $w \in B_\delta \setminus D_\delta$. In addition, from (6.6) and (6.9), we can conclude that there exists $\lambda(\delta)$ such that $\tilde{f}(w) + \lambda \tilde{g}(w)$ points into B_δ^o for all $\lambda < \lambda(\delta)$ and all $w \in D_\delta$. Hence, for sufficiently small λ, $\tilde{f}(w) + \lambda \tilde{g}(w)$ points into B_δ^o for all $w \in B_\delta$. This situation is illustrated in figure 6.11.

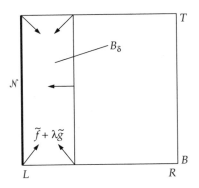

Figure 6.11
Interior-pointing perturbed dynamics

The next step is to apply Brouwer's theorem to the function $\Psi : B_\delta \to B_\delta$ defined, for small enough ϵ, by $\Psi(w) = w + \epsilon(\tilde{f}(w) + \lambda \tilde{g}(w))$, where $0 < \delta < \delta^*$ and $0 < \lambda < \lambda(\delta)$, and where ϵ is chosen sufficiently small to ensure that $\Psi(w) \in B_\delta$ for all $w \in B_\delta$. A fixed point w^* of $\Psi(w)$ is a stationary state of $\dot{w} = \tilde{f}(w) + \lambda \tilde{g}(w)$; any such stationary state is contained in B_δ^o.

As $\lambda \to 0$, proposition 6.1.1 indicates that these stationary states approach $\mathcal{N}$. Because all trajectories through points of B_{δ^*} do not leave B_{δ^*} and nonconverging trajectories are excluded by hypothesis, the basin of attraction of the set of all stationary states of the perturbed dynamics inside B_{δ^*} contains B_{δ^*}, no matter how small λ is. The component $\mathcal{N}$ thus satisfies the criteria for being a potential description of the long-run behavior of the perturbed system.

The dynamics in figure 4.2 are sufficiently simple that the component $\mathcal{N}$ could hardly fail to have well-behaved stability properties, and this is evident without the machinery of the preceding argument. We must now use this argument to extend the result to more complex situations. To do this, we can first note that the argument holds for a broader range of cases than the component $\mathcal{N}$ in the game of figure 4.1. All that is required is a component of equilibria $\mathcal{N}$ that is the face of a state space W and that satisfies (6.6)–(6.7). The result will then hold regardless of the dimension of W.

The next step is to look for cases in which components of equilibria may not be faces of the phase diagram but are structurally equivalent to a face. Let Z be a state space, though not necessarily the state space of figure 4.2. Let $C \subset Z$ be a closed component of stationary states of

$\dot{z} = f(z)$. We need to specify what it means for C to have properties similar to those of $\mathcal{N}$ in the previous discussion, and to identify the implications of this similarity.

Some notation is required. Let the state space be denoted by W as well as Z. Let $\dot{z} = f(z) + \lambda g(z)$ be continuously differentiable on Z. If $\mathcal{N}$ is a face of the state space, recall that $B_\delta(\mathcal{N})$ is the set of points whose distance from $\mathcal{N}$ in the max norm is δ or less. Binmore and Samuelson [27] present the proof of the following.[11]

Proposition 6.2   Let $f + \lambda g$ be continuously differentiable on the state space Z. Let $\mathcal{N}$ be a face of the state space W $(= Z)$. Suppose there exists a differentiable injection $\phi : W \to Z$ with differentiable inverse Φ such that $C = \phi(\mathcal{N})$ and such that the dyamics $\tilde{f}(w) = \Phi'(\phi(w)) f(\phi(w))$ and $\tilde{g}(w) = \Phi'(\phi(w)) g(\phi(w))$, together with the face $\mathcal{N}$ and the state space W, satisfy (6.6)–(6.7). Then there exist $\delta^* > 0$ and $\lambda(\delta) > 0$ for all $0 < \delta < \delta^*$ such that stationary states of $f(z) + \lambda g(z)$ can be found in $\phi(B_\delta^o(\mathcal{N})) \subset Z$ whenever $0 < \delta < \delta^*$ and $0 < \lambda < \lambda(\delta)$. The basin of attraction of the set of such stationary points contains $\phi(B_{\delta*}(\mathcal{N}))$. As $\lambda \to 0$, these stationary points converge to $C \subset \phi(B_\delta(\mathcal{N}))$.

To see why this result holds, notice that the mapping ϕ is a diffeomorphism, meaning that it is a differentiable bijection with a differentiable inverse, between the state space (denoted by W) and a subset of the state space containing the component C. From the preceding analysis of $\mathcal{N}$ and W, we know there exist $\delta^* > 0$ and $\lambda(\delta)$ such that $\tilde{f} + \lambda \tilde{g}$ has a stationary state w^* in $B_\delta^o(\mathcal{N}) \subset W$ for $0 < \delta < \delta^*$ and $0 < \lambda < \lambda(\delta)$, given the condition that $\tilde{f}$ and $\tilde{g}$ satisfy (6.6)–(6.7). But ϕ is differentiable and hence Φ' is nonsingular. This implies that $z^* = \phi(w^*)$ is a stationary state of $f + \lambda g$ contained in $\phi(B_\delta^o(\mathcal{N}))$ and whose basin of attraction contains $\phi(B_{\delta*}(\mathcal{N}))$.

If the conditions of proposition 6.2 are satisfied, how does this imply that the set C is worthy of attention? For any small δ and sufficiently small λ, the system $f + \lambda g$ has stationary states that lie within distance δ of C. In addition, these stationary states have a basin of attraction that contains $\phi(B_{\delta*})$ and hence that does not shrink as λ shrinks. The set C thus provides an approximation of the local limiting behavior of the dynamic $f + \lambda g$ for small λ.

In the two applications of this result that follow, the mapping ϕ will be chosen so that the set $\phi(B_\delta)$ will be the set of points in the basin of

---

11. Φ' is the Jacobian of the differentiable function Φ.

attraction of C that lie with distance δ of C. To say that such a ϕ exists and causes condition (6.6) to hold is then to say that near the component C, the basin of attraction of C is nicely behaved, in the sense that the basin locally looks like the state space, and the trajectories in this basin approach C sufficiently directly, so that on the boundary of $B_\delta(C)$, the learning dynamics lead into $B_\delta(C)$. These are similar to the properties of the dynamics surrounding a hyperbolic sink used to show that a perturbed dynamic must have a nearby sink. The interpretation of ϕ causing condition (6.7) to hold is straightforward, indicating that drift can push the system off the component C, but in so doing must push the system into the basin of attraction of C.

The conditions of proposition 6.2 are sufficient but not necessary for the dynamic $f + \lambda g$ to have interesting stationary states near the component C. In the following two examples a partial converse of this proposition holds, in that the selection dynamic f allows us to readily find a mapping ϕ such that (6.6) is satisfied, and the conclusions of proposition 6.2 hold if and only if g then satisfies (6.7).

The conclusions of proposition 6.2 hold no matter how small λ becomes, that is, no matter how insignificant drift is. Once again we have the point that even very small amounts of drift have a significant effect. The selection dynamics approach zero as we near C because C is a component of stationary states. Even small amounts of drift can then overwhelm the selection process. Proposition 6.2 directs attention to the crucial question: Does this drift tend to push the system toward or away from C? If (6.7) holds, then the drift pushes the system toward C and we have stationary states of the process with drift near C.

6.5   Examples

This section illustrates the effect of drift with two examples. We assume that the selection process is monotonic and regular and that g is completely mixed, or equivalently that it points inward on the boundary of the state space. I find this second assumption especially natural, as the unmodeled forces captured by drift are likely to contain some factors that have nothing to do with payoffs in the game and are capable of introducing any strategy.

Backward Induction

First, return to the Chain-Store Game, whose general form is given by figure 6.1. Let the specific payoffs be as given in figure 1.1. Assessing the impact of drift involves comparing the relative slopes of the learning and drift dynamics, which requires some notation. Let (r, n) denote a point in the unit square, where n is then the proportion of population I playing N (not enter) and r is the proportion of population II playing R (or resisting entry). For a point $z \equiv (r, n) \in Z$, the selection process f then specifies a pair $(\dot{r}(z), \dot{n}(z)) = (f_r(z), f_n(z))$. The component $\mathcal{N}$ of Nash equilibria that are not subgame-perfect is then given by $\{(r, 1) : r \geq \frac{1}{3}\}$.

To investigate the possibility of stationary states near the component $\mathcal{N}$, consider the components of the vector f at states $z \in \mathcal{N}$. Let $f_n(z)/f_r(z)$ be defined as $\lim_{z_k \to z} f_n(z_k)/f_r(z_k)$, where z_k is a sequence of interior states contained in a trajectory approaching z.[12] Because g is assumed to be inward pointing, then $g_n(z) < 0$ for $z \in \mathcal{N}$, and hence $g_n(z)/g_r(z)$ is well defined, though it may again take on the value ∞. Binmore and Samuelson [27] prove

Proposition 6.3 Let g be completely mixed and let f be the monotonic.

(i) Suppose there exists $\theta \geq 1/3$ such that[13]

$$\frac{f_n(\theta, 1)}{f_r(\theta, 1)} < \frac{g_n(\theta, 1)}{g_r(\theta, 1)} < 0. \tag{6.10}$$

Then there exists ϕ satisfying the conditions of proposition 6.2 such that (6.6)–(6.7) are satisfied with $C = \{(r, 1) : r \geq \theta\} \equiv C_\theta$.

(ii) Suppose there is no $\theta \geq 1/3$ such that

$$\frac{f_n(\theta, 1)}{f_r(\theta, 1)} \leq \frac{g_n(\theta, 1)}{g_r(\theta, 1)} \leq 0. \tag{6.11}$$

Then for sufficiently small λ, $f + \lambda g$ has a unique stationary state that converges to $(0, 0)$ as λ converges to zero.

Figure 6.12 illustrates the case in which condition (i) holds, namely, that we can find a subset C_θ of Nash equilibria in which the entrant stays out, and with the property that, on this set, the drift dynamics g point into the basin of attraction of C_θ under the selection process f. If we take the set C_θ to be an interval connecting $(\theta, 1)$ and $(1, 1)$, then the

---

12. Attention is restricted to cases where this limit exists. We must define $f_n(z)/f_r(z)$ as such a limit because z is a stationary state of the dynamics, giving $f_n(z)/f_r(z) = 0/0$. Because f is assumed to be regular, monotonic, and continuously differentiable, a trajectory approaching z exists. In addition, $f_r < 0$ along this trajectory, ensuring that the terms $f_n(z_k)/f_r(z_k)$ are well defined.

13. I adopt the convention here that it is not the case that $\infty < \infty$.

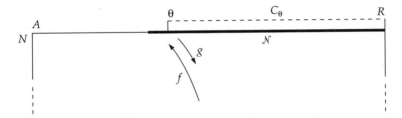

Figure 6.12
Condition (i) of proposition 6.3

key to verifying that drift points into the basin of attraction under the selection process is verifying this property at the endpoint $(\theta, 1)$. The existence of such a set then hinges upon finding a value of θ for which (6.10) holds.

It is important to note that in figure 6.12, C_θ has been chosen to be a *subset* of the Nash equilibrium component $\mathcal{N}$. It will typically be the case that conditions (6.6)–(6.7) can be satisfied not by an entire component of Nash equilibria but by a subset of that component.[14] In the case of the replicator dynamics, the endpoint of the component $\mathcal{N}$, given by $(\frac{1}{3}, 1)$, gives $f_n((\frac{1}{3}, 1))/f_r((\frac{1}{3}, 1)) = 0$, precluding the possibility that (6.10) holds. At this point, interior-pointing drift cannot point into the basin of attraction under the learning dynamics.

If there exists a set θ satisfying (6.10), then the component C_θ has robust stability properties in the perturbed dynamic $f + \lambda g$, even though there are stationary states of the dynamic f, contained in $\mathcal{N} \setminus C_\theta$ that are arbitrarily close to C_θ. The drift g pushes the system from points in C_θ back into the basin of attraction of C_θ and away from points in $\mathcal{N} \setminus C_\theta$. This is where drift plays its essential role.

Proposition 6.3 indicates that when examining the Chain-Store Game, we can simply check the relative slopes of the learning and drift processes on the component of Nash equilibria $\mathcal{N}$. The component $\mathcal{N}$ is worthy of our attention if and only if there are states in $\mathcal{N}$ where the slope of the drift process is flatter than the learning dynamics, in the sense that the drift process points into the basin of attraction, under the dynamics without drift, of a subset of the component of Nash equilibria. "Worthy of attention" means that there are nearby stationary states

---

14. I will often say that "conditions (6.6)–(6.7) are satisfied" rather than the more precise but cumbersome statement that "there exists a diffeomorphism ϕ such that the induced dynamics $\tilde{f}$ and $\tilde{g}$ satisfy (6.6)–(6.7)."

of the process with drift whose basin of attraction does not shrink as the drift level shrinks. For the case of the replicator dynamics and the Ultimatum Minigame, chapter 5 showed that if condition (i) of proposition 6.3 holds, then there is a unique stationary point close to C_θ, which is asymptotically stable.

Comparative Statics

One motivation for an interest in drift is the observation that conventional theoretical predictions often do not match how people (or at least experimental subjects) play games. How does proposition 6.2 help? In particular, what are the potentially testable predictions we could extract from proposition 6.2?

In thinking about testing, it is useful to consider the way that applied work is pursued in conventional economic theory, which consistently uses comparative statics results as a basis for empirical work. Models are used to generate statements about the directions in which variables move in response to changes in observable parameters, while maintaining ceteris paribus assumptions that unobserved parameters do not change.

This suggests that game theorists similarly seek predictions based on comparative statics exercises in which one or more observable parameters are varied while the remaining unobservable parameters are held fixed. If these predictions are violated in the laboratory, then the theory on which they are based must be reconsidered. Notice, however, that this reconsideration may take many forms. The predictions will inescapably depend on maintained hypotheses about unobservables. If a prediction fails, it may then be a reflection of a faulty model or a faulty maintained hypothesis. This is again a familiar characteristic of empirical work.

The payoff function π of a game is the obvious observable, something that can be controlled and varied in an experimental design. In the case of the Chain-Store Game, payoffs are determined by the parameters a, b, c, d, e, and h of figure 6.1. The strategic character of the Chain-Store Game is preserved by subjecting the parameters to the constraints

$$a > e > c \qquad b > d. \tag{6.12}$$

In claiming that payoffs can be manipulated and are given by the parameters a, b, c, d, e, and h, I do not claim that players are motivated only by monetary payoffs. While I think that both game theorists and their critics are often too quick to construct new models of preferences in order to explain unexpected experimental results, I also do not think that monetary payoffs are always the only thing that matters. Instead, some middle ground must be appropriate. But it remains the case that the parameters a, b, c, d, e, and h can be observed and manipulated. Rather than debating whether the forces shaping behavior are better described as preferences or learning, the more useful approach is to derive the implications of the various models and see which is most useful.

The process $\dot{z} = F(z) + \lambda G(z)$ will be treated as unobservable. This is not to say that data cannot be gathered that is relevant to how people learn. The problem is that we do not know how to incorporate these data into the theory in a reliable manner, much less how to make the next step from theory into experiments. For example, it is common in some quarters to insist that if a learning theory is to be useful or convincing, then it must explain the variation in behavior over the rounds of an experiment. I think this insistence is appropriate only if the laboratory experiment provides sufficiently many trials for the learning theory to be applicable. This often fails to be the case, given the necessarily short lengths of many experiments and the emphasis of many learning models on limiting results.

Learning models may capture the process by which players learn to play the games they frequently encounter outside of the laboratory. If the laboratory setting is successful in triggering analogies in the players' thinking to the appropriate games, then we can observe the outcome of this learning process in the experimental behavior. It is this outcome that is addressed by the comparative statics presented here. If the experiment fails in triggering behavior rules among the players that have evolved in the real-life counterpart of the experimental game, perhaps because framing effects direct attention to superficially similar but strategically quite different games, then we can expect to observe very little about the players' learning processes.[15]

A maintained hypothesis about the learning process f, derived as an approximation of $F(z, \pi)$, will be that it is not only regular and mono-

---

15. This emphasis on avoiding framing effects does not imply that we can dispense with the provision of experience in a laboratory setting. It is only by providing such experience that we can collect any hint at all as to how contaminated the results are by framing considerations.

tonic, but also *comparatively monotonic*.[16] To define the latter, denote the ith strategy of player ℓ by $s_{\ell i}$ and let $f_{\ell i}(z, \pi)$ be the coordinate of f corresponding to strategy i for player ℓ. Let $\pi_\ell(s_{\ell i}, z)$ be the average payoff to strategy $s_{\ell i}$ for player ℓ in state z. Now consider two payoff functions $\pi : S \to \mathbb{R}^n$ and $\pi' : S \to \mathbb{R}^n$ and fix a state z. Suppose there exists a strategy $i \in S_\ell$ for player ℓ such that $\pi_h(s_{hj}, s_{-h}) = \pi'_h(s_{hj}, s_{-h})$, for all s_{-h}, whenever $j \neq i$ or $h \neq \ell$, and such that $\pi_\ell(s_{\ell i}, z) > \pi'_\ell(s_{\ell i}, z)$. If $z_{\ell i} > 0$, then it is assumed that $f_{\ell i}(z, \pi) \geq f_{\ell i}(z, \pi')$, and if $z_{\ell j} > 0$ for $j \neq i$, then it is assumed that $f_{\ell j}(z, \pi) \leq f_{\ell j}(z, \pi')$, while f_{hi} for $h \neq \ell$ is unaffected. This assumption ensures that if we fix a state z and then consider a change in the payoffs to player ℓ from strategy i that increases the average payoff of strategy i in state z, then the rate at which strategy i grows increases, and the rates at which other strategies grow for player i decrease.

We know even less about the drift process g, derived as an approximation of $G(z, \pi)$, than about f, and the process of drift clearly goes into the unobservables list. I expect drift to depend upon a host of factors in addition to the payoffs in the game. As a result, drift will often have little to do with the payoffs. In most of what follows, the maintained hypothesis is that g is independent of the payoffs in the game. I refer to this as "the case of exogenous drift."

There remains the question of the initial condition $z(0)$. The comparative statics results derived below are predicated on the assumption that $z(0)$ is independent of payoffs. The predictions will therefore not apply to games for which $z(0)$ turns out to vary significantly with π. Again, we have the point that if framing effects are important, then the theory may have nothing to say.

Applying these considerations leads to the suggestion that experiments compare versions of the Chain-Store Game with varying payoffs. The first task is to determine which payoff configurations are more likely to give stable stationary points near the component $\mathcal{N}$. Binmore and Samuelson [27] prove

Proposition 6.4 Fix payoffs a, b, c, d, e, and h satisfying (6.12) for the Chain-Store Game of figure 6.1. Let the selection dynamic f be monotonic, regular, and comparatively monotonic, and let the drift g be exogenous. If there exists a subset C of the component $\mathcal{N}$ of Nash

---

16. I now replace $F(z)$ by $F(z, \pi)$ to capture the dependence of the selection process on payoffs.

	L	R
T	7, 4	7, 4
M	9, 3	0, 0
B	0, 0	6, 6

Figure 6.13
Dalek Game

equilibria for which a mapping ϕ exists satisfying the conditions of proposition 6.2, then such a subset also exists when either or *e* or *d* is increased or when either *a*, *b*, or *c* is decreased, provided that (6.12) is preserved. The converse can fail in each case.

Chain-Store Game experiments can thus be conducted with varying values of the payoffs *a* and *b*. An outcome consistent with the theory would be the observation of the subgame-perfect equilibrium for large values of *a* and *b* and the Nash, non-subgame-perfect equilibrium for small values of *a* and *b*. Violations of this relationship would challenge the theory. Similar experiments can be done with other combinations of payoffs.

Drift has been taken to be exogenously fixed throughout this exercise, which is often a suitable first approximation, as in biological examples. In the Ultimatum Game, however, we have seen that the drift process may be related to payoffs, in that a population may have a higher drift rate because the payoff differences between its strategies are smaller. The particular drift process examined in chapter 5 reinforces the effects of movements in payoffs, and proposition 6.4 continues to hold for this drift process. An analysis like that leading to proposition 6.4 can be performed for any process of drift, though it will be more complicated and the results will be clearer in some cases than in others, depending upon what is assumed about drift.

Outside Options and Forward Induction

Consider the "Dalek Game" shown in figure 6.13. The extensive form of this game is shown in figure 6.14. Kohlberg and Mertens [135] use this game to illustrate forward induction reasoning.

The Dalek Game has two components of Nash equilibria, including a strict Nash equilibrium given by (M, L) with payoffs $(9, 3)$ and a

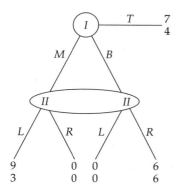

Figure 6.14
Dalek Game, extensive form

component $\mathcal{N}$ of equilibria with payoffs $(7,4)$ in which player I takes the outside option (plays T) and player II plays R with probability at least $\frac{2}{9}$. The former is a (hyperbolic) sink under regular, monotonic dynamics while the stationary states in the latter component are not hyperbolic.

It is common to argue that forward induction restricts attention to the equilibrium (M, L) in this game. For example, we might appeal to the iterated elimination of weakly dominated strategies (cf. Kohlberg and Mertens [135]). B is strictly dominated for player I. Removing B causes R to be weakly dominated for player II, the removal of which causes T to be weakly dominated for player I, leaving (M, L). Alternatively, we could appeal to the never-weak-best-response criterion (Kohlberg and Mertens [135]) or to the forward induction reasoning of van Damme [238, 239] (in an equivalent but different extensive form in which player I first chooses between T and $\{M, B\}$ and then chooses between M and B) or to the normal form variant of this reasoning given in Mailath, Samuelson, and Swinkels [146].

How does this forward induction argument fare in our terms? First, it is straightforward to apply proposition 6.2 to the component $\mathcal{N}$. Letting z_R be the proportion of population II playing strategy R in state z we have

Proposition 6.5 If there is a state $z^* \in \mathcal{N}$ such that $g(z^*)$ points into the interior of the basin of attraction, under the dynamic f, of the set $C = \{z \in \mathcal{N} : z_R \geq z_R^*\}$, then there exists a mapping ϕ satisfying the conditions of proposition 6.2.

In the presence of drift, we thus have good reasons to be interested in Nash equilibria that fail conventional forward induction arguments. However, it would be nice to have a condition that is more revealing than the requirement that "$g(z^*)$ points into the interior of the basin of attraction of the set $C = \{z \in N : z_R \geq z_R^*\}$." The key to finding such a condition is the observation that strategy B is strictly dominated by T for player I in the Dalek Game. Proposition 4.5 shows that any monotonic dynamic will eliminate B. Hence, every trajectory of the dynamics must approach the boundary face spanned by $\{T, M\} \times \{L, R\} \equiv Z_{-B}$. But if attention is restricted to this face, the Dalek Game is a special case of the Chain-Store Game.

Can we use what we know about the Chain-Store Game to identify sufficient conditions for existence of stationary points near the component $\mathcal{N}$ in the Dalek Game? In particular, suppose that condition (i) of proposition 6.3 holds when the dynamics of the Dalek Game are restricted to the face Z_{-B}. Can this fact be used to conclude that the Dalek Game has a stationary point close to $\mathcal{N}$? The answer is yes, as long as drift does not create too powerful a force toward strategy B.

If z is a point in the state space of this game, let z_T, z_B, and z_R denote the probabilities with which strategies T, B and R are played at z, with the residual probabilities being attached to M and L. Similarly, let $f_T(z)$ be the element of the vector f corresponding to T, and so on. Then $\mathcal{N} = \{z : z_T = 1, z_R \geq 2/9\} \subset Z_{-B}$ is the set of states corresponding to the component of Nash equilibria that are not subgame-perfect.

Let $\hat{f}$, $\hat{g}$, and $\hat{f} + \lambda \hat{g}$ be the dynamics f, g and $f + \lambda g$ on Z_{-B} defined by letting $\hat{f} = f$ (this is possible because Z_{-B} is forward-invariant under f) and letting $\hat{g}_T(z) = g_T(z)$ (and hence $\hat{g}_M(z) = -\hat{g}_T(z)$, making Z_{-B} forward-invariant). Call $\hat{f} + \lambda \hat{g}$ and the state space Z_{-B} the "restricted dynamics." Binmore and Samuelson [27] prove

Proposition 6.6 Suppose that for all $\delta > 0$, there is a sufficiently small λ such that the restricted dynamics have a sink (saddle) [source] within δ of $\mathcal{N}$. Then for all $\delta > 0$, there is a function $k(\lambda) : \mathbb{R} \to \mathbb{R}$ and a sufficiently small λ such that if $sup_{z \in Z}|g_B(z)| < k(\lambda)$, then the unrestricted dynamics $f + \lambda g$ have a sink (saddle) [saddle] within δ of $\mathcal{N}$ in the Dalek Game.

To interpret this result, note that proposition 6.3 provides sufficient conditions for the restricted dynamics to have stationary points close to the component of Nash equilibria $\mathcal{N}$ in the state space Z_{-B}, as well as sufficient conditions for no such stationary point to exist. Proposition 6.6 indicates that if the former sufficient conditions are met, then we need look no further. As long as drift does not introduce the strategy B into the game with too much force, the unrestricted dynamics in the Dalek Game also have a stationary point near $\mathcal{N}$.

This result holds because the learning dynamic f pushes the system toward the face Z_{-B}. If perturbations never introduced strategy B into the population, then the system would eventually be governed by the forces that appear in this face; hence the system would eventually have behavior specified by the restricted dynamics on Z_{-B}. Proposition 6.3 could then be applied to gain insight into this behavior. Proposition 6.6 indicates that this continues to hold in the presence of drift that also introduces the strategy B into the population, as long as it does not do so too vigorously.[17]

Comparative Statics

Turning now to comparative statics considerations similar to those of the Chain-Store Game, we could construct a general form of the Dalek Game, the first two rows of which would match the Chain-Store Game of figure 6.1 and the final row of which would correspond to a strictly dominated strategy for player I. From propositions 6.4 and 6.6, we will then have results that are consistent with the model if we observe outcomes near the Nash equilibrium that fails forward induction, in which the outside option is taken, for small values of a, b, and c and large values of d and e. Violations of this pattern would again challenge the theory. One implication is that we should observe Nash equilibria that fail forward induction when the value to player II of taking the outside option is relatively large.

In light of this, it is interesting to note that Balkenborg [5], when conducting experiments with a version of the Dalek Game, finds that the outside option is virtually always chosen. This provides the beginnings of a comparative statics exercise, but additional insight into the model requires additional experiments with different payoffs.

---

17. Notice that if the conditions of proposition 6.6 are met, a source in the restricted dynamics corresponds to a saddle in the unrestricted dynamics. This reflects the fact that f must reduce the proportion of the population playing B and hence must approach the face Z_{-B}.

Binmore et al. [22] examine a related game in which players I and II have ten dollars to divide. Player I demands a share of the ten dollars; player II simultaneously either demands a share of the ten dollars or elects to take an outside option. If player II takes the outside option, then player II receives a payoff of α, where $0 < \alpha < 10$, while player I receives nothing. If player II makes a demand rather than taking the outside option and if the two players' demands are incompatible, meaning that the total of their demands exceeds ten dollars, then each receives nothing. If the demands sum to less than ten dollars, then each player receives his demand as well as half of the remaining surplus.

This game has many Nash equilibria. Pure-strategy Nash equilibria include efficient equilibria, in which player II forgoes the outside option and players I and II make demands that exactly exhaust the ten dollars and allocate at least α to player II, as well as inefficient equilibria, in which player II opts out and player I demands more than $10 - \alpha$. There are also mixed-strategy Nash equilibria, all of which are inefficient. Notice that efficiency requires player II to forgo the outside option, even if α is large, though player II must be compensated for doing so with a payoff at least as large as α.

Forward induction arguments do not exclude any of these equilibria, although it is common to focus on the efficient equilibria of such a game, sometimes with differing opinions as to how player II should be compensated for surrendering the outside option. What if we examine the long-run outcomes of an evolutionary model? Binmore et al. [22] show that, in the presence of drift, the long-run outcomes in this game tend to involve divisions of the surplus that are not too extreme. The more important drift is, the stronger this centralizing tendency, that is, the larger the set of asymmetric divisions of the surplus that do not appear as long-run outcomes. This centralizing tendency appears even though the drift itself is not biased toward symmetric divisions, and for much the same reason that perturbations in the Nash [165] demand game push players toward the Nash bargaining solution.[18] If player II's outside option is not too attractive, this centralizing tendency poses no problems for efficiency, but if the outside option is sufficiently attractive, then the centralizing tendency can make it more

---

18. Similar forces appear in Young's evolutionary model of play in the Nash demand game [251].

attractive for player II to take the outside option than divide the surplus with player I.

The implication of learning with drift, then, is that the outside option is more likely to be chosen, and an inefficient result more likely to appear, when player II's outside option is relatively lucrative. The experimental results of Binmore et al. [22] show that the outside option is often taken, and is taken much more often when it is more lucrative.

6.6   Discussion

The central idea behind this chapter is simple: for a model to be successful, it must include important factors and exclude unimportant ones. But how do we know what is important and what is not? In the case of evolutionary games, the model itself provides the answers. If the model produces stationary states that are not hyperbolic and do not occur in components satisfying some variation of asymptotic stability, then it has excluded important factors.

The factors to be added to the model can have a significant impact on the behavior of the dynamic system, but they also may be very small in magnitude. It is presumably because they are small that they are excluded from the model in the first analysis. How can a model whose behavior is shaped by arbitrarily small factors be of any use in applications? One conclusion of this chapter is that, while the factors themselves may be small, their existence can nevertheless be used to derive comparative statics results that do not depend upon observing arbitrarily small magnitudes.

7   Noise

We begin our investigation of ultralong-run models. As in chapter 3, our point of departure is a model of the stochastic process by which agents adjust their strategies. We shall maintain the convention of previous chapters that this is a Markov process, where a state is a specification of how many of the agents in a finite population are playing each of the strategies available in the underlying game. The Aspiration and Imitation model of chapter 3 is an example of such a Markov model, but less structure will be assumed in this chapter.[1]

We shall be focusing on the stationary distribution of the Markov process. Two familiar forces act on agents' strategies. One is selection. In an economic context, selection occurs via a process of learning; in a biological context, this would be a process of natural selection driven by different strategies leading to different rates of reproduction. The second process is mutation. In a biological context, these are literally mutations, consisting of random alterations in genetic codes; in economic contexts, it is again common to use the term *mutation*, where this refers to an agent randomly switching to a new strategy. We might think of this as an agent experimenting with a new strategy or being replaced by a new agent unfamiliar with the game who initially chooses a strategy at random.

What is the difference between selection and mutation? As in chapter 6, we can think of selection as capturing the important forces affecting strategy choices, while mutation is a residual, capturing whatever has been excluded when modeling selection. As a result, we can expect selection to occur with a higher probability or at a faster rate than mutation, and also to be related to the payoffs in the game. This link

---

1. This chapter is taken from Samuelson [200, 201].

between payoffs and selection is often captured by building the selection mechanism around a tendency for agents to switch toward a best reply to the current state. Such a tendency may take the form of a deterministic selection mechanism in which every agent switches to a best reply to the current state in every period, or it may take the form of a stochastic process in which it is simply more likely that agents switch to a best reply, when they do happen to switch, than the reverse. The relative importance of selection is captured by focusing on the case where the probability of a mutation is very small. To give meaning to the phrase "very small," we shall follow the common practice of examining the limiting outcome as the probability of a mutation approaches zero.

Unlike the selection mechanism, there is little reason to believe that mutations have a particularly strong connection with the payoffs of the game. Thus mutations are conventionally modeled as being completely random, in the sense that they may switch a player to any of the existing strategies and are unaffected by payoffs.

The innovation of Foster and Young [79], developed further by Kandori, Mailath, and Rob [128] and by Young [250], was to show that even arbitrarily improbable mutations can play a crucial role in shaping the stationary distribution of the model. They showed that techniques developed by Freidlin and Wentzell [81] could be used to derive particularly strong conclusions about the limiting case as the probability of a mutation goes to zero. Chapter 6 used "noise" to refer to relatively improbable, stochastic events such as mutations. The message of Young and of Kandori, Mailath, and Rob is simple: noise matters.

This chapter presents a straightforward generalization of the model of Kandori, Mailath, and Rob [128] beyond the two-player 2×2 games and the specific learning and mutation processes they consider, as well as the tools needed to characterize the stationary distribution of this model. These lead to the first result concerning equilibrium selection, namely, that we cannot always expect weakly dominated strategies to be excluded from the stationary distribution. The finding of previous chapters, that dominated strategies are not eliminated by the deterministic differential equations used to approximate the long-run behavior of the strategy adjustment process, is then not an artifact of examining insufficiently long time periods. Dominated strategies persist even in the stationary distribution.

7.1 The Model

Selection

Consider a finite, normal-form game, denoted $G = (\eta, S, \pi)$, where η is a finite set of players, $S = \prod_{\ell \in \eta} S_\ell$ is the joint strategy set, and $\pi = (\pi_1, \ldots, \pi_n)$ is the payoff function.

Assume that G is an asymmetric game and there exist n populations, one for each player in the game. The model is easily adapted to handle the common alternative in which G is symmetric and is played by agents drawn from a single population.

Each population is composed of N identical agents, where N is finite. The agents from these populations are randomly matched to play game G in each period $t \in \{0, 1, 2, \ldots\}$. Chapter 3 mentioned ways of making this "random matching" assumption precise. The key characteristic in each case is that an agent's expected payoff equals the payoff the agent would receive against a mixed strategy whose probabilities are given by the fractions in which the various pure strategies appear in the opposing populations.

At each time t, each agent will be characterized by a pure strategy. The state x of the system at time t identifies, for each population $\ell \in \eta$ and each strategy $i \in S_\ell$, the number of agents of population ℓ playing pure strategy i (denoted $x_{\ell i}$ or $x_{\ell i}(t)$, but with t generally suppressed). Let x, y, and z refer to states and let X denote the set of all possible states.

In each period, after agents have played the game, they potentially switch strategies. Let the process by which agents switch between strategies be called "selection" or a "selection mechanism." A selection mechanism is then a transition matrix defined on the state space X that identifies, for each state the system might occupy at time t, the probability that the system moves to each of the various states in X in period $t + 1$. Hence a selection mechanism is a transition matrix, denoted by F, that identifies the transition probability F_{xy} for all states x and y in X.

For example, suppose that in each period t, each agent in each population takes an independent draw from a Bernoulli trial. With probability $1 - \mu$, this draw produces the outcome "do not learn" and the agent does not change strategies. With probability μ, this draw produces the outcome "learn." In this case, the agent switches to a best response to the period t actions of the agents in other populations (if there is more than one best response, the agent switches to each of them with positive probability). From this model of individual agent behavior, the probabilities F_{xy} that describe the aggregate behavior of the system can be calculated. I refer to these as the "best-response dynamics with inertia." If $\mu = 1$, then we have the deterministic best-response dynamics in which every agent switches to a best reply in every period.

The best-response dynamics with inertia may be applicable in situations where calculating optimal actions is costly, difficult, or time-consuming, so that agents will usually simply repeat the actions they have taken in the past. Occasionally, however, agents in such an environment will assess their strategies. When doing so, they may be able to collect considerable information, perhaps by observing and consulting others, thereby attaining knowledge of the current state. The best-response dynamics with inertia assume that agents then choose a best reply to this state.

The selection mechanism is a Markov process with a transition matrix F and state space X. A stationary distribution, denoted $\gamma = (\gamma_1, \ldots, \gamma_{|X|})$, is a probability measure on X that satisfies $\gamma F = \gamma$. It would be most convenient if the Markov process defined by F had a unique stationary distribution, but we cannot expect this to be the case. Consider, for example, a two-player 2×2 normal-form game with two strict Nash equilibria. Let the selection mechanism be given by the best-response dynamics with inertia. Then there will be at least two stationary distributions, with each attaching unitary probability to a state that corresponds to one of the game's two strict Nash equilibria. The stationary distribution to which the system converges will depend upon initial conditions.

Noise

The key insight of Foster and Young [79] is that adding noise to an evolutionary system, in the form of stochastic mutations, can yield a unique stationary distribution.

The process of mutation is assumed to be independent across agents. Consider an agent from population ℓ playing strategy $i \in S_\ell$ at the end of period t (i.e., after the agents have been matched, played the game, and possibly switched strategies via the selection mechanism). With probability $1 - \lambda$, for $\lambda \in (0, 1)$, no mutation occurs on this agent and the agent continues to play strategy i. With probability λ, this agent undergoes a mutation, in which case the agent switches to each strategy

$j \in S_\ell$ with the exogenously fixed probability $g_\ell(i, j) > 0$. This muta-tion process bears a superficial resemblance to the selection mechanism given by the best-response dynamics with inertia, with the crucial dif-ference that the mutations need not bear any relationship to payoffs and are "completely mixed," meaning that a mutation may switch an agent to *any* strategy.

The results will not depend on the specification of the function g as long as it is retains the property of being completely mixed and as long as g is independent of λ, so that the effect of a mutation is in-dependent of the probability of a mutation. This assumption could be relaxed, and g allowed to depend on λ, as long as the relative prob-abilities of the strategies to which a mutation might switch an agent remain nonzero and bounded as λ approaches zero. This assumption on ratios of mutation-induced transition probabilities would also al-low a more general formulation in which the effect of a mutation on an agent varied across states. For examinations of how other mutation processes might affect the results, see Blume [36] and Bergin and Lip-man [13]. Bergin and Lipman, for example, show that virtually any outcome can be obtained, in the limit as the probability of a muta-tion approaches zero, if in the process the relative probabilities of the strategies to which a mutation switches an agent can approach zero or infinity.

Let G_{yz} denote the probability that mutations move the system from state y to state z. The conjunction of the selection and mutation mecha-nisms gives us an overall transition matrix between states, denoted by Γ, where

$$\Gamma_{xz} = \sum_{y \in X} F_{xy} G_{yz}. \tag{7.1}$$

Notice that $G_{yz} > 0$ for all $y, z \in X$ and hence $\Gamma_{xz} > 0$ for all $x, z \in X$.

The state space X and the transition matrix Γ again define a Markov process. We are interested in the stationary distributions of this process. Chapter 3 observed that the following is standard for a transition ma-trix such as Γ that has only positive elements (see Billingsley [17], Seneta [214], or Kemeny and Snell [132, theorems 4.1.4, 4.1.6, and 4.2.1]):

Proposition 7.1

(i) The Markov process defined by Γ has a unique stationary distribu-tion, that is, a probability measure γ^* on X such that $\gamma^* \Gamma = \gamma^*$.

(ii) The relative frequencies with which the states in X are realized along a sample path of the process approach γ^* almost surely as $t \to \infty$.

(iii) For any initial distribution over states γ_0, we have $\lim_{t \to \infty} \gamma_0 \Gamma^t = \gamma^*$.

Hence, independently of the initial condition, the distribution of the population is asymptotically given by γ^*. This stationary distribution can be interpreted both as the proportion of time that any realization of the process spends in each state and as the probability that the process will be in each state at any (sufficiently distant) future time.

For many, the attraction of this result lies in the stationary distribution of the system being independent of the initial conditions. Alternatively, we say that history does not matter. This greatly simplifies our investigation of the system.

7.2 Limit Distributions

The transition matrix Γ and the stationary distribution γ^* depend on the mutation rate λ, and will sometimes be written as $\Gamma(\lambda)$ and $\gamma^*(\lambda)$. We shall consider the case of *rare* mutations, and hence the limit of the distribution $\gamma^*(\lambda)$ as λ approaches zero, denoted by $\gamma^{**} = \lim_{\lambda \to 0} \gamma^*(\lambda)$. This is the distribution examined by Foster and Young [79], Young [250], and Kandori, Mailath, and Rob [128]. Foster and Young refer to γ^{**} as the "stochastically stable distribution," while Kandori, Mailath, and Rob refer to an equilibrium as a "long-run equilibrium" if this distribution puts a probability mass of one on the equilibrium. I shall refer to it as the "limit distribution" or "limiting distribution."

Counting Mutations

The basic tool used to investigate the limit distribution is a result from Freidlin and Wentzell [81] that was applied in models of evolutionary games by Kandori, Mailath, and Rob [128] and by Young [250]. Consider the following class of directed graphs on the state space X:

Definition 7.1 Let x be a state in X. Then an *x-tree* on X is a tree with each state in X as a node and with x with its initial node, that is, a x-tree is a binary relation on X with the following properties:

(i) For all $y \in X$ with $y \neq x$, there is one and only one element of the form (y, z) in the relation.

(ii) For every $y \neq x$, the relation contains a sequence of the form $(y, z^1), (z^1, z^2), \ldots, (z^m, x)$.

(iii) The relation contains no element of the form (x, y).

If we think of an element of the relation (y, z) as denoting an arrow from y to z, then an x-tree is a tree or collection of arrows such that each element other than x is the initial point of one and only one arrow and there is a path of arrows from each element of $X \setminus \{x\}$ to x. Notice that there are no cycles in such a tree. For each $x \in X$, let h_x denote an x-tree and define the set of x-trees by H_x. It will often be convenient to denote the element (y, z) of an x-tree as $(y \to z)$.

We can now construct the following vector. For a fixed state x and for each x-tree, associate with each (y, z) in the tree the transition probability from y to z, or Γ_{yz}. Then take the product of these probabilities for the tree, and sum these products over the set of all x-trees. This gives a number of the form

$$\zeta_x(\lambda) = \sum_{h_x \in H_x} \prod_{(y,z) \in h_x} \Gamma_{yz} = \sum_{h_x \in H_x} \prod_{(y,z) \in h_x} \sum_{w \in X} F_{yw} G_{wz}. \tag{7.2}$$

We can collect these numbers to construct a vector of the form $\zeta(\lambda) = (\zeta_1, \ldots, \zeta_{|X|})$. Then the result of Freidlin and Wentzell [81, lemma 3.1, p. 177] is

Proposition 7.2 $\gamma^*(\lambda)$ is proportional to $\zeta(\lambda)$.

Freidlin and Wentzell's proof is surprisingly straightforward, involving some simple algebraic manipulations of the condition that $\gamma^*(\lambda)$ is a stationary distribution (i.e, the condition $\gamma^*(\lambda)\Gamma(\lambda) = \gamma^*(\lambda)$).

Proposition 7.2 allows us to derive some properties of the limit distribution from (7.2). First, we can calculate G_{wz}, which identifies the probability that mutations take the system from state w to z. Let M_{wz} identify the various combinations of mutations that can move the state from w to z. Then define $\psi(\ell, m_{wz})$ to be the total number of mutations required of agents in population ℓ under mutation combination $m_{wz} \in M_{wz}$. Let $C(\ell, m_{wz})$ be the binomial coefficient corresponding to the total number of ways that these population ℓ mutants can be drawn from population ℓ given state y and let $G(\ell, m_{wz})$ be the product of the probabilities $g_\ell(.,.)$ for the mutations required of population ℓ under m_{wz}. Then we have

$$G_{wz} = \sum_{m_{wz} \in M_{wz}} \prod_{\ell \in \eta} C(\ell, m_{wz}) G(\ell, m_{wz}) \lambda^{\psi(\ell, m_{wz})} (1 - \lambda)^{N - \psi(\ell, m_{wz})}, \tag{7.3}$$

and from (7.2), we have

$$\zeta_x = \sum_{h_x \in H_x} \prod_{(y,z) \in h_x} \sum_{w \in X} F_{yw} \sum_{M_{wz}} \prod_{\ell \in \eta}$$

$$\times C(\ell, m_{wz}) G(\ell, m_{wz}) \lambda^{\psi(\ell, m_{wz})} (1 - \lambda)^{N - \psi(\ell, m_{wz})}. \tag{7.4}$$

The key property of this equation is that the only terms that depend on λ are those in which λ explicitly appears.

Second, let $P_x = \{y \in X : F_{xy} > 0\}$ and

$$\kappa(x, y) = \min_{z \in P_x} \left[\min_{m_{zy} \in M_{zy}} \sum_{\ell \in \eta} \psi(\ell, m_{zy}) \right]. \tag{7.5}$$

Then we can interpret $\kappa(x, y)$ as the least number of mutations required to get from state x to y, in the sense there is positive probability that starting from state x, the selection mechanism will move the system to a state z from which $\kappa(x, y)$ mutations can move the system to state y (with a similar statement holding for no number less than $\kappa(x, y)$).

Section 7.5 proves the following:[2]

Proposition 7.3

(i) A limit distribution γ^{**} exists and is unique.

(ii) The states appearing in the limit distribution are those which can be reached with the fewest mutations. In particular, $\gamma^{**}(x) > 0$ if and only if x satisfies

$$x \in \arg\min_X \left(\min_{h_x \in H_x} \sum_{(x,y) \in h_x} \kappa(x, y) \right). \tag{7.6}$$

If we interpret the term in parentheses in (7.6) as the cost, in terms of mutations, of linking all other states to the state x, we then have the states easiest to reach from other states appearing in the limit distribution, with "easiest" interpreted as requiring the fewest mutations.

Proposition 7.3 indicates that in order to obtain results on the limit distribution γ^{**}, we can look for counting arguments, where the counting concerns the least number of mutations, that is, the highest power attached to $(1 - \lambda)$, required to accomplish the transitions contained in an x-graph. The fewer the mutations, the more likely x is to appear in the limit distribution, in the sense that only mutation-minimizing

---

2. This result (proposition 7.3) is standard. For example, see theorem 1 of Kandori, Mailath, and Rob [128] or theorem 2 of Young [250].

states will appear. Notice that the only relevant property of the selection probabilities F_{xy} in this calculation is whether they are zero (yielding $\kappa(x, y) > 0$) or positive (yielding $\kappa(x, y) = 0$), with the magnitude of positive F_{xy} being irrelevant. This ability to replace calculation with counting and to restrict attention to the two-member partition of the unit interval into zero and nonzero elements is a great simplification and the gain achieved by working with the limiting case as $\lambda \to 0$. Hence I shall say that the transition $(x \to y)$ "requires $\kappa(x, y)$ mutations" and, similarly, that the transition $(x \to y)$ "does not require a mutation" if and only if $F_{xy} > 0$.

Absorbing Sets

A nonempty set of states $A \subset X$ is *absorbing* if it is a minimal set with respect to the property of being closed under the selection mechanism. Hence there is zero probability that the selection mechanism can cause the process to exit A (though mutations may still move the system out of an absorbing set). In addition, the minimality requirement ensures that there is a positive probability of moving from any state in A to any other state in A in a finite number of time periods. If this were not the case, a strict subset of A could be found that is closed under the selection mechanism, and A would not be absorbing. An absorbing set may contain many states or may contain only a single state. In the latter case, I shall refer to the state as simply an "absorbing state." Because the state space is finite and closed under the selection mechanism, it is immediate that at least one absorbing set exists, and there may be more than one absorbing set.

The *basin of attraction* of an absorbing set is the set of states from which there is positive probability that the selection mechanism (without mutations) moves the system to the absorbing set in a finite number of time periods. Let $B(A)$ and $B(x)$ denote the basin of attraction of an absorbing set A or an absorbing state x. The union of the basins of attraction of the absorbing sets in X must equal X. If the selection mechanism F is deterministic, meaning that only the probabilities zero and one appear in the transition matrix, then the basins of attraction of two absorbing sets must be disjoint. If probabilities other than zero and one appear in F, then the basins of attraction of absorbing sets A and A' may intersect, meaning that there are states from which the selection mechanism may lead to A but may also lead to A'.

Proposition 7.3 leads immediately to the following (proved in section 7.5):

Proposition 7.4

(i) State $x \in X$ appears in the support of the limit distribution only if x is contained in an absorbing set A of the selection mechanism.

(ii) If one state in an absorbing set A of the selection mechanism is contained in the support of the limit distribution, so are all other states in A.

It is a standard result that for a Markov process Γ, the stationary distribution of Γ attaches probability only to states contained in absorbing sets of Γ. Proposition 7.4 indicates that as the probability of a mutation becomes small, the stationary distribution of Γ becomes concentrated on absorbing sets of the selection mechanism F.[3] To see why this holds, we need only note that the selection mechanism leads from any state not contained in an absorbing set to an absorbing set, without requiring any mutations; only with the help of a mutation can the system leave an absorbing set. The number of mutations required in an x-tree that links the remaining states to state x is thus lower if x is a member of an absorbing set than if it is not. The result then follows from proposition 7.3. The proof in section 7.5 makes this intuition precise.

Proposition 7.4 indicates that adding mutations to the selection process F provides a means of choosing between the stationary distributions of the Markov process F. In particular, F may have many stationary distributions. For every absorbing set of F, there is a stationary distribution whose support coincides with that absorbing set.[4] For any λ, however, the process Γ has a unique absorbing set consisting of the entire state space and has a unique stationary distribution. As $\lambda \to 0$, this stationary distribution converges to one of the stationary distributions of F.

To illustrate propositions 7.3 and 7.4, let us apply the argument behind the result of Kandori, Mailath, and Rob [128], that the limit distribution contains only the risk-dominant equilibrium in symmetric 2×2 games with two strict Nash equilibria, to the Stag-Hunt Game shown in figure 2.9. The two strict Nash equilibria of this game consist of (X, X) and (Y, Y). In the terms of Harsanyi and Selten [116], the former equilibrium is payoff-dominant while the latter is risk-dominant. As do

---

3. Gilboa and Samet [101] introduce the concept of an absorbent stable set, which is analogous to the concept of an absorbing set.

4. We can create other stationary distributions by taking combinations of these.

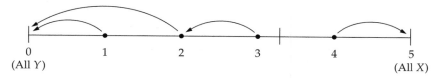

Figure 7.1
Phase diagram for best-response dynamics, Stag-Hunt Game

Kandori, Mailath, and Rob, let this game be played by a *single* population of N agents. The state space is then the set $\{0, 1, \ldots, N\}$, where a
state identifies the number of agents playing strategy X.

Following Kandori, Mailath, and Rob [128], let the selection mechanism F be given by the best-response dynamics, so that in each period
every agent switches to a best response to the distribution of strategies
played by that agent's opponents in the previous period. Figure 7.1
shows the phase diagram for the Stag-Hunt Game under the deterministic best-reply dynamics for the case in which there are five agents.[5]
There are two absorbing sets of the selection mechanism, each of which
is a singleton, consisting of the state 0 and the state 5, which correspond to the game's two strict Nash equilibria. The mixed-strategy
equilibrium is given by $\frac{2}{3}X + \frac{1}{3}Y$. The relative position of this point
is marked on the state space in figure 7.1; it does not correspond to
a state, and hence cannot be an absorbing set. It is important to note,
however, that even if the number of agents in the population were chosen so that we could exactly duplicate the probabilities of the mixed-
strategy equilibrium in population proportions, the mixed-strategy
equilibrium would not be an absorbing state. This is because in single-
population models, each agent samples opponents from a slightly different distribution.[6]

---

5. These transitions are easily computed from the rule that all agents switch to a best
reply to the current state. Notice that if the current state is 3, so that two agents play
Y and three play X, then Y agents face opponents in proportion $\frac{3}{4}X + \frac{1}{4}Y$ and find X a
best reply while X agents face $\frac{1}{2}X + \frac{1}{2}Y$ and find Y a best reply, switching the system to
state 2.

6. To see this, suppose we have a population of 300 agents and assign 200 to X and 100 to
Y. Then an X-agent sees a population of opponents consisting of 199 X-agents and 100 Y-
agents, making Y a best reply. Similarly, agents currently playing Y find X a best reply. If,
instead of working with a single population, we allow each player to be represented by
a separate population, then each agent can face population frequencies of strategies that
match the probabilities of a mixed-strategy equilibrium, and hence the mixed-strategy
equilibrium can be an absorbing state.

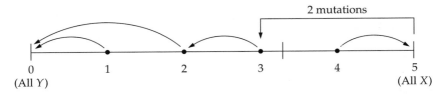

Figure 7.2
Mutation-minimizing 0-tree h_0^*, Stag-Hunt Game

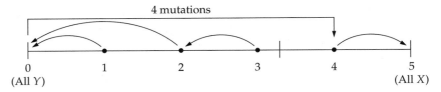

Figure 7.3
Mutation-minimizing 5-tree h_5^*, Stag-Hunt Game

Figures 7.2 and 7.3 show the mutation-minimizing 0-tree and 5-tree, denoted by h_0^* and h_5^*, along with an identification of how many mutations each transition requires. These are the trees that potentially achieve the minimum in parenthesis in (7.6). Fewer mutations are required in h_0^*, which is to say, state 0 satisfies (7.6), while state 5 does not. The limit distribution then places a probability of one on state 0.

It is no coincidence that state 0 is the risk-dominant equilibrium. In a 2×2 symmetric game, figure 7.1 shows that the key calculation is the number of mutations required to make a transition from one of the equilibria to the basin of attraction of the other. The equilibrium with the highest such number will be the only one that appears in the limit distribution. But this will be the equilibrium with the largest basin of attraction, which by definition is the risk-dominant equilibrium.[7]

7. Similar ideas are used in a much different setting by Glance and Huberman [105] to examine the repeated Prisoners' Dilemma. They examine a model in which a group of n players repeatedly meets to play an n-player Prisoners' Dilemma. Players are independently, randomly selected to revise their strategies. When doing so, each player chooses an action that maximizes the player's payoff from the remainder of the game. To evaluate this payoff, each player holds an expectation of how the choices of future players are related to his current choice. Finally, the system is perturbed by random fluctuations in choices. In the most interesting case, expectations are constructed so that for sufficiently long horizons, the proportion of players cooperating in each period is close to one of two levels, a high level and a low level. These are "locally stable," in the sense

Proposition 7.3 tells us about the support of the limit distribution but tells us nothing about the relative probabilities attached to the states in this support. In the case of the Stag-Hunt Game, we have been lucky: the support of the limit distribution is a single state and there can then be no issue of relative probabilities. In other cases, the support may contain absorbing sets that are not singletons or may contain multiple absorbing sets. Much can still be learned, however, simply from knowing the support of the limit distribution; for example, the support may contain many states, but each of these may correspond to the same outcome of the game.

The key factor in the Stag-Hunt Game was the number of mutations required to make a transition from absorbing sets to the basins of attraction of other absorbing sets. We can obtain a more general result concerning the importance of such transitions by pruning our x-trees to eliminate many of the transitions that do not require a mutation.

Definition 7.2 Let A be an absorbing set. An A-*graph* on X given selection mechanism F is a binary relation h_A on X such that

(i) for each state x, there is at most one transition of the form (x, y);

(ii) for every absorbing set $A' \neq A$ there is a sequence of transitions (x^i, x^{i+1}), $i = 1, \ldots n$, with $x^1 \in A'$ and x^n in the basin of attraction of A.

Notice that an A-graph need not contain every element of X nor every element of A as nodes. In addition, an A-graph need not be connected. We could do yet more pruning, but only at the cost of a more complicated definition.

Consider an absorbing set A and a state $x \in A$. Then every x-tree is also an A-graph. This ensures that if we count the number of mutations in a mutation-minimizing A-graph, it cannot exceed the number of mutations in a mutation-minimizing x-tree. On the other hand, there

---

that small random fluctuations pushing the system away from one of these points simply prompt convergence back to the point in question. Larger fluctuations can switch the system from one locally stable point to another, however, and the key to the ultimate behavior of the system is the relative probability of fluctuations switching the system between these two states. This is analogous to the mutation-counting arguments of Kandori, Mailath, and Rob [128]. The models differ in that the Kandori, Mailath, and Rob agents are completely myopic, choosing best responses for the stage game whenever they revise strategies, while the Glance and Huberman agents recognize that they are in a repeated interaction with the same players and evaluate payoffs accordingly. The latter convention is essential to Glance and Huberman's ability to get outcomes other than constant defection from the Prisoner's Dilemma.

are many A-graphs that are not x-trees because they are missing transitions. Notice, however, that any minimal A-graph can be converted into an x-tree for some $x \in A$ by adding transitions that do not require mutations.[8] This then gives us the converse, that the number of mutations in a mutation-minimizing A-graph cannot exceed the number in a mutation-minimizing x-tree for $x \in A$, from which we immediately have

Proposition 7.5   State x appears in the limit distribution if and only if x is contained in an absorbing set A for which

$$A = \arg\min_{A \in \mathcal{A}} \left(\min_{h_A \in H_A} \sum_{(x,y) \in h_A} \kappa(x,y) \right), \tag{7.7}$$

where h_A is an A-graph, H_A is the set of A-graphs, and $\mathcal{A}$ is the set of absorbing sets.[9]

Nash Equilibria

Without making assumptions about the selection mechanism F, we have no reason to believe that the outcome of the model will bear any relationship to familiar equilibrium concepts. This section investigates the consequences of a common assumption, namely, that agents do not switch away from best responses and do not play inferior responses eternally. To formalize this, let $\pi_\ell(i,x)$ be the expected payoff to an agent in population ℓ playing strategy $i \in S_\ell$, given that x is the current state.[10] Then the system is *Nash-compatible* if, for any state $x \in X$,

$$[\exists \ell \in \eta, \exists i, j \in S_\ell : x_{\ell i} > 0, \pi_\ell(i,x) < \pi_\ell(j,x)] \Leftrightarrow [\exists y \in X, y \neq x : F_{xy} > 0]. \tag{7.8}$$

Statement (7.8) indicates that there is positive probability that the state of the system changes if and only if some agent is not playing a best

---

8. This holds because each state z that is not the origin of a transition in an A-graph is either contained in the basin of attraction of A or is contained in the basin of attraction of an absorbing set so that has a state y as the origin of a transition, and transitions can then be added linking state z to x, or z to y, without requiring mutations.

9. Because h_x is an x-tree and h_A an A-graph, some room for ambiguity arises as to what h_x or h_A means when A is a singleton absorbing set containing the element x. The meaning will be clear from the context.

10. This calculation is well defined because x includes a specification of the strategies played by agents in populations other than ℓ, and the matching process ensures that the expected payoff to an agent in population ℓ matches the expected payoff from playing a single agent whose *mixed* strategy duplicates these population proportions.

reply. Although this is a common assumption, it is not immediately clear what ensures that the system will work so flawlessly. Chapter 9 explores the implications of relaxing this assumption in a particular setting. I shall abuse notation somewhat by saying that a *state* is a Nash equilibrium if each agent in the state is playing a best response to the distribution of strategies in the other populations. Hence state x is a Nash equilibrium if and only if, for all $\ell \in \eta$ and any strategy $i \in S_\ell$,

$$[x_{\ell i} > 0] \Rightarrow [\pi_\ell(i,x) \geq \pi_\ell(j,x) \ \forall j \in S_\ell]. \tag{7.9}$$

From (7.8), we immediately have

Proposition 7.6 Let the selection mechanism be Nash-compatible. Then the state x is a Nash equilibrium if and only if x is a singleton absorbing set.

Therefore, if the absorbing sets constituting the support of the limit distribution are all singletons, then the support consists of Nash equilibria.

Proposition 7.6 is another version of the "folk theorem" of evolutionary game theory that "stability implies Nash," with the requirement that the limit distribution contain only singleton absorbing sets supplying the stability condition. It stops short of simply asserting that the limit distribution consists only of Nash equilibria because in general we cannot exclude nonsingleton absorbing sets from the limit distribution. This should come as no surprise. Many evolutionary models yield the result that outcomes will be Nash equilibria if they exhibit some stability property, but conditions ensuring convergence to stable outcomes are much more elusive.[11]

Recurrent Sets

The basic task of characterizing the limit distribution of the Markov process has been replaced with that of characterizing the *support* of the limit distribution, in the hopes that there are enough interesting cases in which the support alone provides sufficient information that we need look no further. How do we characterize the support? We can

---

11. Young's assumption that the game is weakly acyclic [250] is motivated by the need to exclude nonsingleton absorbing sets. Crawford [66], Marimon and McGrattan [147], and Milgrom and Roberts [155, 156] examine conditions under which certain dynamic systems converge.

always return to (7.6) to answer this question, but it would be helpful to have more convenient methods at our disposal. Proposition 7.4 indicates that the support of the limit distribution is a union of absorbing sets of the selection process, and we thus immediately know the support for those cases in which there is only one absorbing set. However, we are interested in cases where there are multiple equilibria and hence multiple absorbing sets.

This section develops a necessary condition for an absorbing set to be in the support of the limit distribution. Techniques for characterizing stationary distributions are also developed by Ellison [76].

The first step is to examine the possible effects of mutations.

Definition 7.3   States x and y are *adjacent* if one mutation can change the state from x to y.

If a single mutation can change the state from x to y, then it can also accomplish the reverse transition, making adjacency an equivalence relation. I will refer to the set of states adjacent to state x as the "single-mutation neighborhood" of x and will denote this set of states by $M(x)$.

Definition 7.4   A collection of absorbing sets R is *recurrent* if it is a minimal collection of absorbing sets with the property that there do not exist absorbing sets $A \in R$ and $A' \notin R$ such that $M(A) \cap B(A') \neq \emptyset$.

Hence a recurrent set R is a collection of absorbing sets with the property that it is impossible for a single mutation, followed by selection dynamics, to lead to an absorbing set not contained in R. Instead, the selection dynamics that follow a mutation must lead the system back to an absorbing set in R, though this need not be the absorbing set containing the state in which the initial mutation occurred. The minimality requirement ensures that R is cyclic in the sense that, for any two absorbing sets A^1 and A^n in R, we can find a sequence of absorbing sets $A^2, \ldots, A^{n-1}$ in R with $M(A^i) \cap B(A^{i+1}) \neq \emptyset$ for $i = 1, \ldots, n-1$. Hence we can move throughout the absorbing sets in R by transitions that include a single mutation followed by selection dynamics.

It is clear from definition 7.4 that every recurrent set is a collection of absorbing sets, but not every absorbing set is contained in a recurrent set. It will often conserve on notation to think of a recurrent set R as the set of states contained in the various absorbing sets that constitute R. It will be clear from the context whether R is taken to be a set of absorbing sets or the set of states contained in these absorbing sets. From the definition of a recurrent set, we immediately have

Lemma 7.1   At least one recurrent set exists. Recurrent sets are disjoint.

The basic result is now that the limit distribution contains only absorbing sets that are contained in recurrent sets, and includes all of the absorbing sets in any recurrent set that contributes to the limit distribution. The limit distribution is thus a union of recurrent sets. Section 7.5 proves

Proposition 7.7   Let state x be contained in the support of the limit distribution. Then

(i) x is contained in a recurrent set R;

(ii) all states $y \in R$ are contained in the support of the limit distribution.

A necessary condition for being contained in the limit distribution is then that a state be contained in a set that is robust to single mutations, which is to say, the selection dynamics that follow a single mutation must at least eventually lead back to the set.

Where we encounter multiple recurrent sets and know only that at least one of them appears in the support of the limit distribution, there is no substitute for returning to proposition 7.3 and either doing the complete mutation-counting analysis or looking for further necessary conditions that might allow us to discriminate between the recurrent sets. Where we encounter a unique recurrent set, but one so large and containing states corresponding to such a variety of outcomes as to be uninformative, there is no substitute for seeking information on the shape of the limit distribution on its support. In many cases, however, we will find a unique recurrent set with much to be learned simply by knowing that the support of the limit distribution is confined to that recurrent set.

The results that emerge in the presence of a unique, "well-behaved" recurrent set are more robust than many mutation-counting arguments. We may be uncomfortable with results driven by the relative likelihoods of such highly improbable events (as λ gets small) as the simultaneous occurrence of enough mutations to move the system from one absorbing set to the basin or attraction of another. For example, we may hesitate to eliminate an absorbing set A from consideration on the strength of the argument that the $10,001$ (highly improbable) mutations needed to get from absorbing set A to A' are much less likely than the $10,000$ mutations needed to get from A' to A. In contrast, results based solely on the composition of recurrent sets are results driven by the fact that it takes only a single mutation to escape from a set that is not recurrent, while it takes more than a single mutation to escape from a recurrent set. I regard comparisons of this type as the most robust to emerge from the model and regard games in which such comparisons are useful as the natural games to examine with the model.[12]

7.3 Alternative Best Replies

The tools developed in the previous section can now be applied to questions of equilibrium selection. Particular attention will be devoted to Nash equilibria with alternative best replies, that is, Nash equilibria in which at least one agent has alternative strategies that yield payoffs equal to those of the equilibrium strategy.

Alternative best replies are a natural application of these tools. As previous chapters have noted, many equilibrium refinement questions arise out of the presence of alternative best replies. More importantly, the mutation-counting arguments that arise in connection with alternative best replies often involve simply checking whether the equilibria in question are contained in a recurrent set; hence single-mutation arguments suffice to exclude such equilibria from consideration.

The following will help characterize recurrent sets:

Definition 7.5   A set of states $X' \subset X$ is a *mutation-connected component of Nash equilibria* if X' is minimal with respect to the properties that every element of X' is a Nash equilibrium and no Nash equilibrium is adjacent to any element of X'.

A mutation-connected component is thus a maximal collection of Nash equilibria with the property that one can move from any state in the component to any other state in the component, without leaving the component, by moving between adjacent states.

A mutation-connected component may appear to be simply a component of Nash equilibria, translated from the joint strategy set of the game into the state space, but this need not be the case. To illustrate

---

12. A complete consideration of "robustness" must include a careful examination of the issue of waiting times. In particular, dissatisfaction arises with results driven by large numbers of mutations because they can require very long periods of time before the stationary distribution is an applicable description of the process. Chapter 9 offers some brief remarks on waiting times.

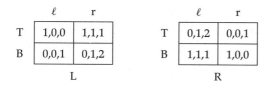

	ℓ	r
T	1,0,0	1,1,1
B	0,0,1	0,1,2

L

	ℓ	r
T	0,1,2	0,0,1
B	1,1,1	1,0,0

R

Figure 7.4
Game with mutation-connected components

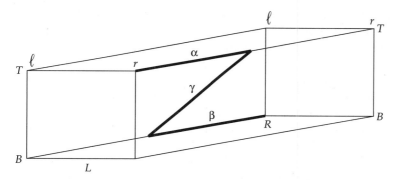

Figure 7.5
Mutation-connected components

the difference, consider the game shown in figure 7.4. Player *I* chooses rows, player *II* chooses columns, and player *III* chooses matrices. There is a single component of Nash equilibria in this game, consisting of three connected line segments in the strategy space, namely, α, β, and γ:

Segment α: Player *I* plays *T*, player *II* plays *r*, and player *III* plays any mixture in which the probability attached to *L* is at least 0.5.

Segment β: Player *I* plays *B*, player *II* plays *l*, and player *III* plays any mixture in which *R* receives probability at least 0.5.

Segment γ: Player *III* attaches probability $\frac{1}{2}$ to both *L* and *R* while players *I* and *II* play any mixture in which the probabilities attached to *T* and *l* sum to unity, as must the probabilities attached to *B* and *r*.

These equilibria are illustrated in figure 7.5.

In the state space, this component of Nash equilibria yields a collection of singleton absorbing sets. However, while segments α and β are each (separately) mutation-connected components, segment γ is not contained in a mutation-connected component. In particular, a movement between two "neighboring" absorbing sets in segments α and β requires only one mutation, switching a population-III agent between L and R or vice versa. A movement between two neighboring states in segment γ requires two mutations, one from population I and one from population II.

For the purposes of our analysis, mutation-connected components are the relevant object. For example, let the learning dynamics in figure 7.4 be given by the best-response dynamics. Then segments α and β are in recurrent sets, but there is no recurrent set containing γ. To verify this, we first note that beginning with a state in segment γ, a single mutation, causing an agent in population III to switch from L to R (or R to L) yields a state where B and l (or T and r) are best replies, which then yields a switch to a state where all agents in populations I and II play B or l (or T or r), contained in segment β (or α). If, however, we begin with a state in segment α or β, no single mutation can yield a state in the interior of segment γ (because the actions of at least two agents must change, one from population I and one from population II); each state in the interior of segment γ has only itself as its basin of attraction (because all populations mix at such states and at no state outside of γ are all populations indifferent, so the best-response dynamics, from any state not in γ, switch at least one population to a pure strategy). This establishes that segments α and β can be contained in recurrent sets, but γ cannot be. This argument is easily extended to show that there is a unique recurrent set consisting of the union of α and β.

The definitions of a mutation-connected component and recurrent set ensure that if one state of a mutation-connected component is contained in a recurrent set, then so are all states. From these definitions, we immediately have

Proposition 7.8 Let the selection mechanism be Nash-compatible. If x is a Nash equilibrium contained in the support of the limit distribution, then the entire mutation-connected component of Nash equilibria containing x is contained in the support of the limit distribution.

Six examples will serve to illustrate proposition 7.8. In the first five, the selection mechanism is assumed to be the best-response dynamics with inertia; in the final, slightly more complicated example, it will be convenient to switch to another form of the best-response dynamics.

For the first example, consider the game shown in figure 4.1, where strategy L strictly dominates R for player II. There is a unique component of Nash equilibria in which player II plays L and player I chooses any mixture between T and B. There is similarly a unique mutation-connected component in the state space, consisting of all those states in which all agents in population II play L, namely, the unique recurrent set. The support of the limit distribution then consists of all those states in which player II plays L. In all but one of these states, some agents in Population I play B, and hence play a weakly dominated strategy.

This, the first of several examples in which we find that dominated strategies need not be excluded from the limit distribution, raises three questions. First, how does strategy B survive? The answer is familiar from chapter 4. T dominates B, but only weakly so, with T earning a strictly higher payoff than B as long as some agents in population II play R. In addition, L strictly dominates R, and agents in population II are thus switching to L. We then have a race between agents in population II switching to L and agents in population I switching to T. If the former switch sufficiently quickly, then the pressure against B is eliminated and an outcome appears in which the dominated strategy B persists.

Second, why do mutations not eliminate strategy B? Considerations from chapter 4 again reappear. By continually introducing R into the population, mutations should create weak but persistent pressure against B. In particular, the selection dynamics near the component of equilibria are always pushing population I away from B and toward T. Given that mutations constantly push us away from this component and bring these dynamics into play, we should then expect the limit distribution at least to concentrate its probability on states in which the proportion of agents playing T is high. Notice, however, that at the same time these mutations introduce a pressure against B arising out of the appearance of R, they also introduce pressure toward B in the form of mutations that switch agents in population I to strategy B. If we simply left mutations in population I to operate unchecked, the result would be a force toward equal numbers of agents playing T and B. The net effect of these two forces is then unclear. Samuelson [201] shows that we will obtain outcomes in which the dominated strategy B is eliminated if we are willing to examine the limiting case of an arbitrarily large population and if, at states where almost all agents play L (formally, at states where all but one agent play L), the probability the selection mechanism switches an agent from B to T becomes arbitrarily large, relative to the probability an agent switches from R to L, as the population grows.

The palatability of this last condition depends upon our view of the selection mechanism. As the population becomes large, the payoff difference between strategies T and B, at states where all but one agent play L, goes to zero. If we are to obtain an outcome in which dominated strategies are eliminated, the probability that agents switch from B to T must not go to zero. Hence the selection mechanism must exhibit a discontinuity, with a zero probability of switching from B to T when the payoff difference is zero (if Nash compatibility is to be preserved) while even an arbitrarily small payoff difference yields a switching probability that is bounded away from zero. I have referred to selection mechanisms with this property as "*ordinal* selection mechanisms" (Samuelson [198]) because strategy switches appear to be driven by the relative payoff rankings of strategies rather than the magnitudes of the payoff differences involved. On the other hand, we might expect the probability of switching between strategies to be increasing in the payoff difference between the two strategies, yielding a *cardinal* system. I am inclined toward cardinal mechanisms, an inclination already indulged in chapter 6, where the selection process f was assumed to be continuously differentiable. If we choose cardinal mechanisms, however, we must be willing to live with ultralong-run outcomes in which dominated strategies appear.

Third, why do we care if strategy B survives? The payoffs are the same regardless of whether (T, L) or (B, L) is played. More generally, we would be unconcerned with the details of the limit distribution as long as all the states in its support give the same economic outcome. However, by changing player II's payoffs in figure 4.1, we can obtain a case with the same limit distribution where the outcome does matter.

In making this change, we obtain a game that is not generic even in its extensive form, much less being generic in the space of normal-form payoffs. As noted in chapter 2, my preference is to restrict analysis to generic games. However, as chapter 2 also noted, the space in which the genericity argument is formulated must be carefully chosen. In many cases it is even too demanding to insist that if two outcomes give one player the same payoff, then they give all players the same payoff because the economic outcomes at the two nodes may differ in details relevant to one player and not the other. Hence a belief that analysis should be restricted to generic games does not obviate the need to seriously consider games with ties between payoffs and games with payoff ties for some but not all players.

For the second example, let us turn to the game shown in figure 4.3, where again there is a unique Nash equilibrium in undominated strategies, given by (T, L). The absorbing sets are all singletons and consist of every state in which either all agents in population I play T or all agents in population II play L. These sets together constitute a mutation-connected component and are a recurrent set; all are then contained in the support of the limit distribution. Once again, we limit our attention to Nash equilibria, but dominated strategies may appear. Notice that some coordination is introduced into the limit outcome. Both strategies B and R can appear, but no state appears in which both strategies are played.

For the third example, consider the game shown in figure 6.4. where, as before, the absorbing sets are singletons corresponding to the Nash equilibria of the game and consist of all those states in which all agents in population I play N and at least $\frac{1}{3}$ of the agents in population II play R. These are again contained in a single recurrent set that is the support of the limit distribution.

For the fourth example, we can present a single-population game with features similar to those of figure 6.4. Young and Foster [252] consider a version of the infinitely repeated Prisoners' Dilemma played by three strategies: TIT-FOR-TAT (TFT), COOPERATE (C), and DEFECT (D). Taking the payoffs of the stage game in the Prisoners' Dilemma to be given by figure 1.5 and using the limit of the means to calculate payoffs, the payoff matrix for this three-strategy game is given by figure 7.6. The phase diagram for this game under a regular, monotonic dynamic is given in figure 7.7. There is again a single recurrent set containing all of the game's Nash equilibria, in which no agents play D and at least half of the agents play TFT. This set is the support of the limit distribution under the best-response dynamics with inertia.

For the fifth example, consider the game shown in figure 7.8, where the Nash equilibria consist of two mutation-connected components. Component α contains those states in which all agents in population II play L and all agents in population I play M or T, with at least $\frac{2}{3}$ playing M. Component β consists of a single state corresponding to the Nash equilibrium (B, R), which, because M dominates B, is an equilibrium in dominated strategies; hence every equilibrium in component β involves dominated strategies.

	TFT	C	D
TFT	3, 3	3, 3	1, 1
C	3, 3	3, 3	0, 5
D	1, 1	5, 0	1, 1

Figure 7.6
Repeated Prisoners' Dilemma with three strategies

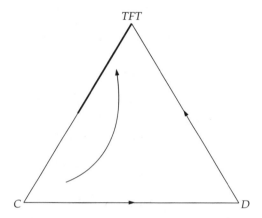

Figure 7.7
Phase diagram for figure 7.6

	L	R
T	1, 0	0, 2
M	1, 2	2, 1
B	0, 1	2, 1

Figure 7.8
Game with a single recurrent set containing dominated strategies

Calculating the recurrent sets for this example requires greater precision in the specification of the best-response dynamics with inertia. What if an agent who receives the learn draw already plays a best-response strategy, but other best responses are also available?

One likely outcome is that the agent will simply retain the current strategy. If this is the case, then the selection mechanism can never cause agents to switch away from strategy M in figure 7.8, and the unique recurrent set consists of component α.

	L	R
T	1, 2	2, 1
M	0, 1	2, 1
B	2, 1	1, 2

Figure 7.9
Game with a single recurrent set containing only dominated strategies

Alternatively, an agent playing a best response may switch to an-other best response. For example, agents may not realize that the cur-rent strategy is a best response, and their analysis of the game may uncover the merits of an alternative best response before those of the current strategy are discovered. In this case, components α and β are contained in a single recurrent set; hence the support of the limit distri-bution consists of both components.[13]

I suspect that players generally require some evidence that an alter-native strategy is better than their current one before switching, and hence can generally be expected to retain best responses. The follow-ing chapter assumes that agents never abandon a best response. Such flawless optimization may be too much to ask, however, and chap-ter 9 examines a case in which agents sometimes mistakenly switch to an inferior reply. The game in figure 7.8 shows that the role of dominated strategies in the limiting outcome can be sensitive to such details.

For our sixth and final example, consider the game shown in fig-ure 7.9. Assume now that the selection mechanism is given by the best-response dynamics, in which every agent switches to a best response in every period. Agents currently playing best responses retain them, and agents not playing best responses, but with multiple best responses available, choose from the latter randomly. Overall, the rigidity of the best-response dynamics makes this example quite contrived, and it

---

13. In particular, the single-mutation neighborhood $M(\beta)$ contains a state in which all agents play B or R except for a single population-I agent who has been switched to M by a mutation. Strategy L is then a best response for agents in population II, so with posi-tive probability the best-response dynamics with inertia cause player-II agents to switch to L and then player-I agents to switch to M, yielding a state in component α. Hence $M(\beta)$ intersects $B(\alpha)$. Now consider the state in component α in which $\frac{1}{3}$ of the agents in population I play T. A single mutation, switching another agent in population I to T, now causes R to be a best response for population II. With positive probability, the se-lection mechanism now switches all population-II agents to R and then all population-I agents to B, giving a state in component β. Hence $M(\alpha)$ intersects $B(\beta)$, and α and β are in the same recurrent set.

provides an extreme outcome: every state in the support of the limit distribution involves dominated strategies. However unreasonable or unlikely such an outcome is, it effectively exposes the mechanics of the model.

Two absorbing sets exist here, each of which is a singleton. Let these absorbing states be denoted by x and y, where state x corresponds to the Nash equilibrium (M, R), which involves the dominated strategy M. State y then corresponds to the mixed equilibrium in which agents in population I are evenly divided between T and B, while those in population II are evenly divided between L and R. To show that the unique recurrent set contains only set x, we can first note that the basin of attraction of state y can only contain states in which player II is indifferent between L and R and attaches an equal proportion of the population to each of these strategies. This in turn is a feature of the well-known difficulty of having best-response dynamics converge to a mixed-strategy Nash equilibrium. However, the best-response dynamics, beginning from any state that can be reached from x by a single mutation, cannot reach a state in which L and R give equal payoffs and are played in equal proportions. The basin of attraction of y is thus not a single mutation away from x.[14] As a special case of the observation that every state except y is in the basin of attraction of x, a single mutation suffices to transform y into a state in the basin of attraction of x.[15] The result is a limit distribution in which only equilibria involving dominated strategies are played.

7.4 Trembles

The previous section shows that dominated strategies can survive in the limit distribution. Chapter 4 has already touched on the differ-

---

14. Any mutation in population I transforms state x into a state in which either L or R is a strict best response for population II. A mutation in population II makes either T or B a strict best response, and once all population-I agents have switched to this best response, either L or R will be a best response for player II. In either case, we reach a state in which all population-II agents play the same strategy. But then population II must thereafter always be monomorphic. In each period, either all population-II agents play a (possibly weak) best response and retain their strategy, or all play an inferior response and switch to the other strategy. This precludes convergence to state y.

15. For example, let this single mutation switch an agent in population II from R to L. Then all agents in population I switch to B, at which point all agents in population II switch to R. Strategies M and T are now both unplayed best responses for player I, allowing some probability that all agents in population I switch to M.

	TL	TR	BL	BR
TL	1	0	0	0
TR	1	0	0	0
BL	0	0	1	0
BR	1	0	0	0

Figure 7.10
Selection mechanism

ences between perturbations to the process by which agents select their strategies, such as the mutations that appear in this chapter, and trembles in the implementation of a strategy. To explore the difference between mutations and trembles in the current context, I add trembles to the model.[16]

Let us begin with an example built around the game of figure 4.1. To make explicit the calculations that characterize the limit distribution, we must work with a simple state space. We therefore assume that each of the two populations consists of a *single* agent. There are then four states, consisting of $\{TL, TR, BL, BR\}$, where the first element of a state specifies player I's choice and the second element specifies player II's choice. Let λ be the probability an agent undergoes a mutation that switches the agent's strategy, and let λ be constant across agents and states.

The selection mechanism without trembles will be constructed to create the strongest possible pressure against the dominated strategy B consistent with payoff-driven strategy choices. We shall therefore assume that if player II plays R in period t, then player II switches to L in period $t + 1$. Player I switches strategies if and only if a payoff of zero is realized, in which case player I switches to T. The transition matrix for the selection mechanism is then given in figure 7.10. Notice that states TL and BL are adjacent singleton absorbing sets and hence together constitute the unique recurrent set. In the absence of trembles, the limit distribution will thus include TL and BL.

Now let us consider trembles. Before the play of the game in each period, each player takes a draw from a Bernoulli trial that with probability δ produces the outcome "tremble" and with probability $1 - \delta$

---

16. Canning [58, 59] examines evolutionary models with trembling hand trembles but without mutations.

	TL	TR	BL	BR
TL	1	0	0	0
TR	1	0	0	0
BL	δ	0	$1-\delta$	0
BR	1	0	0	0

Figure 7.11
Selection mechanism with trembles

produces no tremble. In the latter case, the agent plays her current strategy; in the former, the agent's strategy is chosen from a tremble distribution that depends upon the agent's current strategy.

Given the selection mechanism, trembles on the part of player I can do nothing to speed the elimination of strategy B. Accordingly, let us assume that trembles never distort player I's choices, so that the tremble distributions for player I simply put unitary mass on player I's current strategy. Player II's tremble distribution when playing R similarly puts unitary mass on R because having player II play L instead of R can never increase the pressure to play B. However, when player II chooses L, the result is to play L with probability $1 - \delta$ and to play R with probability δ. These trembles ensure that T receives an expected payoff higher than B in every state. Together, the trembles and selection mechanism appear to present the best case possible for the elimination of B.

The selection mechanism with trembles is shown in figure 7.11. There is now a unique absorbing set consisting of the state TL, and the limit distribution will place probability only on TL. Trembles thus have the effect of excluding dominated strategies from the limit.

We are interested, however, not in trembles that occur with some fixed probability δ but in trembles that can be taken to be arbitrarily small, just as trembling-hand perfect equilibria are obtained by taking the limit as trembles approach zero. We are then interested in two limits, with both the probability of a tremble and the probability of a mutation approaching zero. Once again, the order of these limits is an issue. To address this, let $\delta(\lambda)$ identify the probability of a tremble as a function of the probability of a mutation, where $\lim_{\lambda \to 0} \delta(\lambda) = 0$. The complete transition matrix, with both trembles and mutations, is then given in figure 7.12.

	TL	TR	BL	BR
TL	$(1-\lambda)^2$	$(1-\lambda)\lambda$	$(1-\lambda)\lambda$	λ^2
TR	$(1-\lambda)^2$	$(1-\lambda)\lambda$	$(1-\lambda)\lambda$	λ^2
BL	$(1-\delta(\lambda))(1-\lambda)\lambda$ $+\delta(\lambda)(1-\lambda)^2$	$(1-\delta(\lambda))\lambda^2$ $+\delta(\lambda)(1-\lambda)\lambda$	$(1-\delta(\lambda))(1-\lambda)^2$ $+\delta(\lambda)(1-\lambda)\lambda$	$(1-\delta(\lambda))(1-\lambda)\lambda$ $+\delta(\lambda)\lambda^2$
BR	$(1-\lambda)^2$	$(1-\lambda)\lambda$	$(1-\lambda)\lambda$	λ^2

Figure 7.12
Transitions with both trembles and mutations

We can now examine the x-trees associated with states TL and BL. For TL, the TL-tree that maximizes the product of transition probabilities consists of the three transitions $(TR \to TL)$, $(BL \to TL)$, and $(BR \to TL)$, for a product of transition probabilities of

$$T_{TL} = [(1-\lambda)^2]^2[(1-\delta(\lambda))(1-\lambda)\lambda + \delta(\lambda)(1-\lambda)^2]. \qquad (7.10)$$

For BL, the BL-tree that maximizes the product of transition probabilities consists of the three transitions $(TL \to BL)$, $(TR \to TL)$, and $(BR \to TL)$, for a product of transition probabilities of

$$T_{BL} = [(1-\lambda)^2]^2[(1-\lambda)\lambda]. \qquad (7.11)$$

In the limit, as $\lambda \to 0$ and hence $\delta(\lambda) \to 0$, the state BL will be excluded from the stationary distribution if and only if

$$\lim_{\lambda \to 0} \frac{T_{BL}}{T_{TL}} = 0. \qquad (7.12)$$

From (7.10)–(7.11), we see that (7.12) will hold if and only if

$$\lim_{\lambda \to 0} \frac{1}{(1-\delta(\lambda)) + (\delta(\lambda)/\lambda)(1-\lambda)} = 0. \qquad (7.13)$$

Equation (7.13) in turn will hold if and only if

$$\lim_{\lambda \to 0} \frac{\delta(\lambda)}{\lambda} = \infty. \qquad (7.14)$$

Hence in the limit trembles must become arbitrarily more likely than mutations. This is not surprising. Trembles can alter the support of the limiting outcome only if they save on mutations—if, for example, trembles allow a transition to be made without a mutation that previously required a mutation. But to "save a mutation" means a transition has to occur with a probability an order of magnitude higher than that of a mutation, and hence (7.14) must hold.

This relationship between trembles and mutations goes beyond our simple example.[17] Let $F_{xy}(\delta)$ be the selection mechanism with trembles, where $F_{xy}(\delta) = F_{xy} + O(\delta)$. Let $\gamma^{**}(\lambda, \delta(\lambda))$ be the "limit distribution with trembles," where $\delta(\lambda)$ gives the probability of a tremble as a function of the probability of a mutation, $\gamma^*(\lambda, \delta(\lambda))$ gives the stationary distribution for fixed λ and hence fixed $\delta(\lambda)$, and where $\gamma^{**}(\lambda, \delta(\lambda))$ is the limit of these distributions as $\lambda \to 0$. Let trembles and mutations be independent.

Proposition 7.9 Let x and y be adjacent absorbing states. Let x but not y appear in the support of $\gamma^{**}(\lambda, \delta(\lambda))$, the limit distribution with trembles. Then $\lim_{\lambda \to 0} \delta(\lambda)/\lambda = \infty$.

Proposition 7.9 is proven in section 7.5. The intuition is familiar from our example. Trembles matter because they create new transitions between states and hence save on mutations, allowing some paths in the state space to be traversed with fewer mutations. If it is to make a difference in the support of the limit distribution, however, "saving a mutation" must involve creating a transition probability an order of magnitude higher than the probability of a mutation. Hence trembles must become arbitrarily more likely than mutations in the limit. The "small" trembles of trembling hand perfection may not be as small as they initially appear.

7.5 Appendix: Proofs

Proof of Proposition 7.3
Statement (i) Because $\gamma^*(\lambda)$ is proportional to $\zeta(\lambda)$, it suffices for the existence of $\gamma^{**} = \lim_{\lambda \to 0} \gamma^*(\lambda)$ to show that for all $x, y \in X$, that $\zeta_x(\lambda)/\zeta_y(\lambda)$ has a limit as λ converges to zero. This follows immediately from (7.4), which implies that $\zeta_x(\lambda)/\zeta_y(\lambda)$ is a ratio of polynomials in λ.   □

Statement (ii) From (7.4), ζ_x can be written as the sum over H_x of terms of the form $\alpha\lambda^\beta(1 - \lambda)^{nN-\beta}$, where α and β are constant in λ. From proposition 7.2, we then have $\gamma^{**}(x) > 0$ if and only if the summation over

---

17. The version of the following argument that appeared in Samuelson [201] was unnecessarily complicated and contained errors. I am thankful to Philippe Rivière for bringing this to my attention.

H_x appearing in ζ_x contains a term that maximizes (over the summations over H_y appearing in the expressions for ζ_y, $y \in X$) the exponent attached to $1 - \lambda$. From the definition of $\kappa(x, y)$ given in (7.5), this is the set of x satisfying (7.6).   □

Proof of Proposition 7.4
Statement (i)   Let

$$N(h_x) = \sum_{(y,z) \in h_x} \kappa(y, z). \tag{7.15}$$

We interpret $N(h_x)$ as the minimum number of mutations required to accomplish the transitions in the x-tree h_x. Let $N^*(x) = \min_{h_x \in H_x} N(h_x)$. Then $N^*(x)$ is the value of the minimum contained in parentheses in (7.6). We assume throughout the remainder of this section that all sequences of states are chosen so that their members are unique, without loss of generality.

Now let $x^1 \in X$ not be contained in an absorbing set. Let $h^*_{x^1}$ be an x^1-tree for which $N(h^*_{x^1}) = N^*(x^1)$. There exists a sequence of distinct states $x^2, \ldots, x^n$ such that $F_{x^i x^{i+1}} > 0$ for $i = 1, \ldots, n-1$ and such that x^n is contained in an absorbing set A and is the first such state in the sequence. There also exists a state $y^1 \in A$ such that $h^*_{x^1}$ contains a transition $y^1 \to z$ for some state $z \notin A$; and there exists a sequence of states $y^2, \ldots, y^n$ contained in A with $F_{y^i y^{i+1}} > 0$, $i = 1, \ldots, n-1$ and $F_{y^n x^n} > 0$.

It suffices to show that $N^*(x^n) < N^*(x^1)$ because this will ensure, from (7.6), that x^1 is not contained in the limit distribution. Now remove from $h^*_{x^1}$ each transition of the form (x^i, w^i) and (y^i, z^i), $i = 1, \ldots n$ (i.e., remove all transitions beginning with x_i and y_i, $i = 1, \ldots n$) and add the transitions (x^i, x^{i+1}), $i = 1, \ldots, n-1$, (y^i, y^{i+1}), $i = 1, \ldots, n-1$, and (y^n, x^n). The result is an x^n-tree. None of the added transitions requires a mutation, while the deleted transition (y^1, z) did require a mutation (because y^1 but not z is contained in the absorbing set A). Hence $N^*(x^n) < N^*(x^1)$, as required.   □

Statement (ii)   Let y^1 and y^n be contained in an absorbing set A. We show $N^*(y^n) \leq N^*(y^1)$, which (because y^1 and y^n are arbitrary) suffices for the result. Let $N(h^*_{y^1}) = N^*(y^1)$. Note that A must contain a sequence of states $y^1, \ldots, y^n$ with none of the transitions (y^i, y^{i+1}) requiring a mutation. Then construct a y^n-tree by deleting from $h^*_{y^1}$ each transition of the form (y^i, z^i) and adding (y^i, y^{i+1}), $i = 1, \ldots, n-1$. The result is a y^n-tree h_{y^n} with $N(h_{y^n}) \leq N(h^*_{y^1}) = N^*(y^1)$, giving the result.   □

Proof of Proposition 7.7 For a state x^1, let $S(x^1)$ be the set of states to which the selection mechanism can lead from x^1, so that $x^n \in S(x^1)$ if and only if there is a sequence $x^2, \ldots, x^{n-1}$ with $F_{x^i x^{i+1}} > 0$ for $i = 1, \ldots, n-1$. Notice that if A is an absorbing set and $x^n \in A$, then $x^n \in S(x^1)$ if and only if $x^1 \in B(A)$. We shall similarly define $S(X')$ for a set $X' \subset X$ as the union of the sets $S(x)$ for states $x \in X'$. The statements are proven in reverse order.

Statement (ii) Let A^1 and A^n be absorbing sets contained in a single recurrent set R with states $x^1 \in A^1$ and $x^n \in A^n$ and $x^n \in S(M(x^1))$. We shall show (see the proof of proposition 7.4 for notation):

$$N^*(x^n) \le N^*(x^1). \tag{7.16}$$

This in turn implies that if the elements of one absorbing set in R appear in the support of the limit distribution, so do all elements in all absorbing sets in R and thus all states in R, establishing statement (ii).

To verify (7.16), let $h^*_{x^1}$ be an x-tree with $N(h^*_{x^1}) = N^*(x^1)$. Then there exist $x^2 \in M(x^1)$ and $x^3, \ldots, x^{n-1}$ with $F_{x^i x^{i+1}} > 0$, $i = 2, \ldots, n-1$, and with x^n chosen without loss of generality so that no elements of this sequence are contained in A^n. Also, $h^*_{x^1}$ must contain a transition $(y^1 \to z)$ for some $y^1 \in A^n$ and $z \notin A^n$. Because $y^1 \in A^n$, there exists $y^2, \ldots, y^m$ with $F_{y^m x^n} > 0$ and $F_{y^i y^{i+1}} > 0$, $i = 1, \ldots, m-1$. Then construct an x^n-tree by deleting from $h^*_{x^1}$ the transitions beginning with $x^2, \ldots, x^n$ and $y^1, \ldots, y^m$ and adding

$$x^i \to x^{i+1} \qquad i = 1, \ldots, n-1$$

$$y^i \to y^{i+1} \qquad i = 1, \ldots, m-1$$

$$y^m \to x^n.$$

The only one of the added transitions for which $\kappa(.,.) > 0$ is $(x^1 \to x^2)$, which gives $\kappa(x^1, x^2) = 1$. Since $\kappa(y^1, z) \ge 1$, we have $N(h_{x^n}) \le N(h^*_{x^1}) = N^*(x^1)$, giving (7.16). $\square$

Statement (i) Now let x^1 be such that $S(M(x^1)) \cap A^n \ne \emptyset$ for some absorbing set A^n with x^1 contained in an absorbing set other than A^n. It suffices for statement (i) to show that $N^*(x^n) < N^*(x^1)$ for some $x^n \in A^n$ if A^n is contained in a recurrent set R that does not contain x^1 and $N^*(x^n) \le N^*(x^1)$ otherwise. The proof of the latter is identical to the argument by which we established (7.16); hence we consider the former. We first assume that $A^n = R$, so that the recurrent set R contains just the single absorbing set A^n. Let h^*_{x1} be an x^1-tree with $N(h^*_{x1}) = N^*(x^1)$. Because $S(M(x^1)) \cap A^n \neq \emptyset$, there exists $x^2, \ldots, x^{n-1}$, with $x^2 \in M(x^1)$ and $F_{x^i x^{i+1}} > 0$, $i = 2, \ldots, n-1$. By suitably choosing x^n without loss of generality, we can ensure that x^n is the first element in this sequence contained in A^n. Also, h^*_{x1} contains transitions (y^i, y^{i+1}), $i = 1, \ldots, m$ and (y^m, x^1) with $y^1 \in A^n$ and $y^2 \notin A^n$ and with (y^k, y^{k+1}), the first transition in this sequence after (y^1, y^2) that requires a mutation. (Such a y^k exists because $y^1 \in M(A^n)$ and A^n is recurrent, ensuring that $S(M(A^n)) \cap B(A^1) = \emptyset$.) Furthermore, there is a sequence of states $z^1, \ldots, z^p$ in A^n with $F_{y^1 z^1} > 0$, $F(z^i z^{i+1}) > 0$, $i = 1, \ldots, p-1$, and $F_{z^p x^n} > 0$; as well as a sequence $w^1, \ldots, w^q$ with $F_{y^k w^1} > 0$, $F_{w^i w^{i+1}} > 0$, $i = 1, \ldots, q-1$, and $F_{w^q v^k} > 0$ for some $v^k \in \{z^1, \ldots, z^p, x^n\}$ or some $v^k \in \{x^1, \ldots, x^n\}$ and where no element of $\{w^1, \ldots, w^q\}$ is contained in $\{z^1, \ldots, z^p\}$ or $\{x^1, \ldots, x^n\}$. Then construct h_{x^n} by deleting from h^*_{x1} every transition beginning with $x^1, \ldots, x^n$, y^1, y^k, $z^1, \ldots, z^p$, and $w^1, \ldots, w^q$. Then add the transitions

$$x^i \to x^{i+1} \qquad i = 1, \ldots, n-1$$

$$y^1 \to z^1$$

$$z^i \to z^{i+1} \qquad i = 1, \ldots, p-1$$

$$z^p \to x^n$$

$$y^k \to w^1$$

$$w^i \to w^{i+1} \qquad i = 1, \ldots, q-1$$

$$w^q \to v^k.$$

The result is an x^n-tree. In addition, of the transitions added to h^*_{x1}, only $(x^1 \to x^2)$ requires a mutation, whereas of the deleted transitions, $\kappa(y^1, y^2) \geq 1$ (because $y^1 \in A^n$ and $y^2 \notin A^n$) and $\kappa(y^k, y^{k+1}) \geq 1$. Hence $N(h_{x1}) < N(h_x) = N^*(x)$, giving the result.

Now suppose that A^n is not the only absorbing set in F. Then the proof proceeds as before, except that we cannot ensure that $F_{w^i w^{i+1}} > 0$ for $i = 1, \ldots, q-1$. However, we can ensure that if $F_{w^i w^{i+1}} = 0$, then $w^i \in A^i$ and $w^{i+1} \notin A^i$ for some absorbing set A^i, with $A^i \neq A^n$ but $A^i \in R$. We can also ensure that there is at most one such i for each absorbing set in R. If there exists i with $F_{w^i w^{i+1}} = 0$, find the associated absorbing set A^i. First delete the transitions beginning with the states in $A^i \setminus \{w_1, \ldots, w_q\}$. Then follow the procedure for constructing the tree h_{x^n} as above. Because of the preliminary deletions, the results need not be an x^n-tree. However, at least one of these preliminary deleted transitions must have required a mutation because A^i is absorbing and these transitions are drawn from an x^1-tree for $x^1 \notin A^i$. Hence the addition of the transitions (w^i, w^{i+1}) and the deletion of the transitions beginning with states in $A^i \setminus \{w^1, \ldots, w^q\}$ leave the total mutation count unchanged. Then, for each such A^i, add a sequence of transitions (v^i, v^{i+1}) with the properties that $F_{v^i v^{i+1}} > 0$ for each of these transitions, each element in $A^i \setminus \{w^1, \ldots, w^n\}$ is the origin of one and only one transition, and the result is a path from each element of $A^i \setminus \{w^1, \ldots, w^q\}$ to $\{w^i, \ldots, w^q\}$. Because A^i is absorbing, this is possible. This yields an x_n-tree without introducing any new mutations, which suffices for the result.   □

Proof of Proposition 7.9   Let $\Gamma_{wz}(\lambda, \delta(\lambda))$ be the probability of a transition from w to z, given mutation probability λ and tremble probability δ. Fix δ and let h_x^* be an x-tree, with $N(h_x^*) = N^*(x)$. (See the proof of proposition 7.4 for notation.) Then h_x^* contains a transition $(y \to z)$. Because h_x^* can be converted into a y-tree by deleting $(y \to z)$ and adding $(x \to y)$, x can be contained and y excluded from the support of ζ^{**} only if

$$\lim_{\lambda \to 0} \frac{\Gamma_{yz}(\lambda, \delta(\lambda))}{\Gamma_{xy}(\lambda, \delta(\lambda))} = \infty.$$

The probability of the transition $(x \to y)$ given by $\Gamma_{xy}(\lambda, \delta(\lambda))$ is at least $O(\lambda)$. The probability of a transition $(y \to z)$ given by $\Gamma_{yz}(\lambda, \delta(\lambda))$ is $O(\delta(\lambda)) + O(\lambda)$ because $F_{yz} = 0$. We then have

$$\lim_{\lambda \to 0} \frac{O(\delta(\lambda)) + O(\lambda)}{O(\lambda)} = \infty,$$

which requires $\lim_{\lambda \to 0} \delta(\lambda)/\lambda = \infty$, giving the result.   □

8 Backward and Forward Induction

8.1 Introduction

This chapter puts the tools developed in chapter 7 to work. We shall focus on whether the limit distributions in very simple extensive-form games restrict attention to outcomes that exhibit backward or forward induction properties.[1]

To examine extensive-form games, two modifications in the previous analysis are required.[2] First, the notions of what characterizes a player and a state must be expanded. A player, instead of being described simply by strategy, will now be characterized by a strategy and a conjecture as to the strategies of other players; a state, by a specification of how many agents in each population hold each possible conjecture and strategy.

I think it most natural to assume that players in an extensive-form game can observe the path of play through the game but *cannot* observe what opponents would have chosen at unreached information sets, and hence cannot observe opponents' strategies. To make choices based on their observations of current outcomes, players need some conjectures as to what behavior will appear if these choices cause information sets to be reached that are not included in the current path of play. As a result, a process in which players choose best responses to current *strategies* can require players to know the details of behavior at information sets that cannot be observed under current play. Endowing players with conjectures allows them to choose best responses to observable information and to their beliefs about things they cannot observe.

---

1. This chapter is taken from Nöldeke and Samuelson [168].
2. Canning [58, 59] confronts similar issues in examining evolutionary processes on extensive-form games.

Second, once conjectures are introduced into the model, the evolutionary "folk theorem" that "stability implies Nash" must be reconsidered. Singleton absorbing states will now correspond to self-confirming equilibria rather than to Nash equilibria. Absorbing sets exist in which each agent is playing a best reply to the observed behavior of other agents, but which fail to be Nash equilibria because conjectures about what opponents would do off the equilibrium path are not correct.

Given that we are going to be working with self-confirming equilibria, the notion of backward induction also requires some modification. A self-confirming equilibrium will be said to be "subgame-consistent" if the strategies and conjectures yield a self-confirming equilibrium on every subgame. Notice that this notion of consistency could be applied to any equilibrium concept, and a subgame-perfect equilibrium is simply a Nash equilibrium that is subgame-consistent.

The examination begins with the study of recurrent outcomes. An outcome is *recurrent* if there exists a recurrent set with the property that every state in that set produces the outcome in question. There may be many states producing a given outcome, where these states differ in their specifications of conjectures or behavior off the equilibrium path in ways that do not affect realized behavior. This chapter shows that recurrent outcomes exhibit both backward and forward induction properties. In extensive-form games in which each player moves at most once along any path, every recurrent outcome is a subgame-consistent self-confirming equilibrium outcome. Furthermore, every recurrent outcome must satisfy a forward induction property. In two-player games from our class, this property implies the never-weak-best-response property.

Our analysis cannot halt with this result, however, because recurrent outcomes may fail to exist. Instead, recurrent sets may contain states yielding a variety of outcomes.[3] The chapter continues by considering two simple classes of games.

First, we examine backward induction in games of perfect information in which each player moves at most once along any path. In these games there is a unique recurrent set, which contains the subgame-perfect equilibrium outcome. However, the subgame-perfect equilib-

---

3. It is also possible for recurrent outcomes to exist without the corresponding recurrent set appearing in the support of the limit distribution or constituting the entire support of the limit distribution.

rium outcome will be the only outcome in this set, and hence will be a recurrent outcome, only under stringent conditions. If these conditions are not met, the subgame-perfect equilibrium still appears in the limiting distribution, but is accompanied by other self-confirming equilibria that yield different outcomes.

Next, we examine forward induction in games where player I has a choice between exercising an outside option or playing a normal-form game with player II. If there is *any* strict Nash equilibrium of the subgame yielding a higher payoff to player I than the outside option, then equilibria in which the latter is played cannot appear in the limiting distribution.

8.2   The Model

Consider a game denoted by G, where G is now taken to be a finite, extensive-form game with perfect recall and without moves by nature. Let $\eta = \{1, \cdots, n\}$ denote the set of players and Ξ the set of terminal nodes, or outcomes. An assignment of payoffs to terminal nodes is given by the function $\pi : \Xi \to \mathbb{R}^n$.

Our attention will be restricted to games G satisfying the condition that no path through the game tree intersects two or more information sets of a single player. This is a natural class of games in which to examine subgame perfection because questions of rational behavior after evidence of irrationality do not arise (cf. Binmore [18] and Reny [186]). More important, restricting attention to this class of games allows the specification of straightforward rules for updating strategies and conjectures.[4]

Learning

Given an extensive form game G, assume that, for every player ℓ, there is a finite population consisting of $N > 1$ agents. At each time $t \in \{0, 1, 2, \ldots\}$, every possible combination of agents capable of playing the game meets and plays.

At time t, each agent of each population is described by a *characteristic*, consisting of a pure behavior strategy and a conjecture. A conjecture for an agent specifies, for each population representing an opponent and for each action such an opponent may take at one of his

---

4. See notes 6 and 8 for details.

information sets, the number of agents in the population taking that action.[5] A *state* of the system is a specification of how many agents in each population have each possible characteristic. Let X denote the set of possible states of the system, and let x denote an element of X. Associated with every state is a distribution over terminal nodes, denoted by σ_x, that results from the matching process and play of the game.

The selection mechanism will be given by a form of the best-response dynamics with inertia. Hence, in each period, after agents have been matched, each agent of each population independently takes a random draw from a Bernoulli trial. With probability $1 - \mu \in (0, 1)$, the agent's characteristic does not change. With probability μ, the draw produces the outcome "learn." An agent who learns is able to observe the *outcomes* of all matches of the current round of play. Notice that players cannot observe actions at information sets that are not reached during the course of play.

Given this information, the agent first updates his conjectures to match the observed frequency of actions at all information sets that were reached during period t. Updated conjectures about play at information sets that are reached are uniquely determined and the same for all agents who learn.[6] Conjectures at unreached information sets are unchanged.[7]

Given his new conjecture, the agent updates his behavior strategy. At all information sets where his current action is a best response,

---

5. Notice that these conjectures describe opponents' actions rather than beliefs about nodes in information sets, as in the usual specification of a sequential equilibrium (see Kreps and Wilson [138]).

6. This specification uses the assumptions that every player moves at most once along each path and that learning players observe the outcomes of all matches. In particular, suppose player i's information set h is reached in some match. Whether h is reached depends only on the choices of player i's opponents, so it must be that for all agents from population i, there is a matching that allows their behavior at h to be observed. All learning agents then observe this behavior and update their conjectures to match the actual distribution of actions by player-i agents at information set h. The assumption that learning agents observe the current state is strong, but allows a convenient characterization of conjectures.

7. Because agents update conjectures to match the most recently observed play, an agent's conjecture about another agent is a strategy rather than a probability distribution over strategies. Assuming that learning agents observe the most recent play and change their conjectures to match this play yields a finite state space, which simplifies the analysis.

his action remains unchanged.[8] At all information sets where his current action is not a best response against his conjecture, the agent changes to an action that is chosen according to some probability distribution (which depends only on the agent's information about the current state) that puts positive probability on all actions that are best responses.

In chapter 6 the assumption that the selection mechanism is "smooth" was identified as an important and appealing assumption. The most common specification violating this assumption is best-response dynamics. Best-response dynamics with inertia would satisfy this smoothness assumption if the degree of inertia were related to payoffs, so that agents were less likely to evaluate their strategies and switch to a best response when differences in payoffs were small. The specification of this chapter, however, where the probability of an agent's switching to a best response is μ, regardless of payoffs, is *not* smooth.

This chapter shows that even in the ultralong run, we cannot expect evolutionary models to select subgame-perfect equilibria. Using a best-response-based learning process makes it as difficult as possible to obtain this result because it ensures that the pressure to switch to best responses, which is crucial to obtaining subgame-perfect equilibria, remains strong even as the payoff differences driving this pressure shrink. The result that the subgame-perfect equilibrium is not selected is most powerful in such a setting.

The best-response dynamics with inertia have the property that an agent can hold a conjecture while experiencing outcomes that contradict that conjecture, and yet still not alter the conjecture. Only when the learn draw is next received does an agent check whether her conjectures are consistent with her own play. It may not be unusual for agents to only occasionally evaluate their beliefs by collecting information about the state of the world, but why does an agent not immediately incorporate into her conjectures information provided by the outcome of her own plays? To answer this, I again appeal to the belief

---

8. Again, whether one of the agent's own information sets is reached does not depend on the agent's behavior strategy. Therefore his conjectures and hence action remains unchanged at any information set that was not reached in the previous state. In particular, changes in an agent's actions at an information set h cannot cause that agent to now potentially reach and hence choose a new action at the previously unreached information set h'.

that people are typically involved in many strategic situations, each of which requires them to form conjectures about the situation they face and choose an action based on that conjecture. It is impossible to conduct a full analysis of every choice, and most choices are made on the basis of decision rules that involve little or no analysis. These choices may not be best replies and may be based on conjectures that are inconsistent with observed outcomes or even internally inconsistent. Only when the occasional opportunity is taken to conduct an analysis of a problem can we expect people to form consistent models. These models may subsequently be rendered obsolete by changes in the strategic environment, and the agent will again persist in using a model that is contradicted by her experience, until the next time she brings her reasoning to bear on the problem.

The learning mechanism defines a collection of probabilities of the form F_{xy} for all x and y in X. These probabilities in turn constitute a Markov process on the state space X.

Mutations

Mutations now make their appearance. At the end of each time t, each agent takes another independent draw from a Bernoulli trial. With probability $1 - \lambda \in (0, 1)$, this draw produces no change in this agent's characteristic. With probability λ, this draw produces the outcome "mutate," where the agent changes to a characteristic randomly determined according to a probability distribution that puts positive probability on each of the characteristics possible for this agent.

The learning and mutation processes together yield a Markov process, denoted by Γ, with transition probabilities $\{\Gamma_{xy}\}_{x,y\in X}$. We again have the standard result that the Markov process $\{\Gamma_{xy}\}_{x,y\in X}$ has a unique stationary distribution $\gamma^*(\lambda)$. We are interested in the limit distribution $\gamma^{**} = \lim_{\lambda\to 0} \gamma^*(\lambda)$.

Limit Distribution

The study of the limit distribution begins with a characterization of singleton absorbing sets.

Definition 8.1 A state $x \in X$ is a *self-confirming equilibrium* if each agent's strategy is a best response to that agent's conjecture and if each agent's conjecture about opponents' strategies matches the opponents' choices at any information set that is reached in the play of some match.

This is the analogue of Fudenberg and Levine's notion of a self-confirming equilibrium [91], except that it is defined directly in the state space of the Markov process rather than in the strategy space of the game.[9]

The definition then immediately implies (where absorbing states were introduced in the previous chapter):

Proposition 8.1 State $x \in X$ is an absorbing state if and only if x is a self-confirming equilibrium.

As expected, this establishes equivalence between singleton absorbing sets and self-confirming equilibria rather than Nash equilibria. Limit distributions composed of singleton absorbing sets will then be composed of self-confirming equilibria.[10]

Proposition 7.7 can be adapted to this model to ensure that the limit distribution contains only recurrent sets and all of the absorbing sets in any recurrent set that contributes to the limiting distribution. Then what can be said about recurrent sets? Consider a self-confirming equilibrium. A key observation is that any recurrent set containing this absorbing state will also contain states corresponding to the same path through the extensive-form game, but a variety of out-of-equilibrium behavior and out-of-equilibrium conjectures. Some additional terminology is helpful in making this precise. Given a self-confirming equilibrium outcome, *player ℓ can force entry into a subgame $G(a_\ell)$* if there

---

9. See Kalai and Lehrer [126] for a similar equilibrium concept. There may be self-confirming equilibria in the game G that cannot be supported as self-confirming equilibria in the state space of the Markov process Γ because the former require mixed strategies that cannot be duplicated as population proportions in the finite populations playing game G. A similar difficulty arises when considering subgame-perfect equilibria. In the following analysis we shall not distinguish between equilibria in the game G and in the state space of the Markov process. This is appropriate if the population size is such that we *can* achieve any desired equilibrium on every subgame in the state space. Similar results can be obtained for pure-strategy versions of the equilibrium concepts without this assumption.

10. Nöldeke and Samuelson [167] provide an example of a simple signaling game in which a self-confirming equilibrium that is not a Nash equilibrium appears in the limiting distribution of the evolutionary process, ensuring that the consideration of self-confirming equilibria that are not Nash equilibria is necessary.

exists an information set h belonging to player ℓ that is reached in at least some match during the play of the self-confirming equilibrium and if there also exists an action a_ℓ for player ℓ at h that is not chosen by any player-ℓ agent and is such that the decision node resulting from the choice of a_ℓ is the root of subgame $G(a_\ell)$. We then have

Proposition 8.2

(i) Let $x \in X$ be an absorbing state. Let x' differ from x only in actions prescribed at information sets that are not reached in any matchings in state x (i.e., every agent in x can be paired with a distinct agent in x' such that the two agents hold identical conjectures at all information sets and play identical actions at all information sets reached in any matching under x). Then x' is also an absorbing state and a recurrent set that contains x also contains x'.

(ii) Suppose that, given an absorbing state x, player ℓ can force entry into a subgame $G(a_\ell)$. If x' differs from x only in the conjectures that agents from populations other than ℓ hold over choices at information sets in $G(a_\ell)$, then a recurrent set that contains x also contains x'.

The proof is given in section 8.7. Part (i) of proposition 8.2 indicates that if a self-confirming equilibrium is contained in an absorbing set, then so is any state that can be obtained by altering actions at unreached information sets, while holding conjectures fixed. That conjectures are fixed and actions are affected at unreached information sets ensures that any such new state is also a self-confirming equilibrium and can be reached from the original state by a series of mutations. This implies that the states are contained in the same recurrent set. Part (ii) indicates that the absorbing set will also include any states that can be reached by altering the conjecture of a player about an unreached subgame into which the player cannot force entry. The implication of this result is that the evolutionary process is not effective at imposing discipline at unreached information sets. If a self-confirming equilibrium appears in a recurrent set, so does any state that can be reached by perturbing actions at unreached information sets.

8.3   Recurrent Outcomes

We now examine recurrent sets that are composed of singleton absorbing sets. The only hope for single-valued equilibrium predictions clearly lies with recurrent sets that are composed of absorbing states, all of which correspond to the same self-confirming equilibrium outcome. We call the outcome involved a "recurrent outcome:"[11]

Definition 8.2   An outcome σ^* is *recurrent* if there exists a recurrent set R such that for all $x \in R$, it is the case that $\sigma_x = \sigma^*$.

Studying such outcomes allows us to characterize the limiting distribution when the latter is "nicely behaved" and also provides some clue as to when it will be so behaved.

We can first establish a backward induction property for recurrent outcomes. Let x be a self-confirming equilibrium. Then x is a *subgame-consistent self-confirming equilibrium* if, for any subgame, the actions and conjectures specified by x for this subgame are a self-confirming equilibrium.[12] We then have[13]

Proposition 8.3   Let G^* be an extensive-form game with each player moving at most once along each path. Then every recurrent outcome is a subgame-consistent self-confirming equilibrium outcome.

The proof is given in section 8.7. The intuition is that if the absorbing states in a recurrent set giving outcome σ^* do not correspond to a subgame-consistent self-confirming equilibrium, then there must exist a subgame not reached in outcome σ^* that can be entered by some player ℓ and that has the property that every self-confirming equilibrium of this subgame gives ℓ a higher payoff than does σ^*. In addition, proposition 8.2 ensures that the recurrent set R contains states in which the actions on this subgame correspond to a self-confirming equilibrium, though agents' conjectures do not match those actions (otherwise, agent ℓ would not be playing a best reply, and the state would not be an absorbing state). Now let a mutation cause a player-ℓ agent to enter this subgame, and let all agents learn. Whatever else happens, the result will be a self-confirming equilibrium in the subgame, and subsequent learning cannot change this behavior. The learning process

---

11. It is clear that a nonsingleton absorbing set must contain states corresponding to different outcomes, that is, must contain states x and x' such that $\sigma_x \neq \sigma_{x'}$ because only then can the learning dynamic cause the system to cycle between the states in the absorbing set.

12. Because attention is restricted to games with the property that every path of play intersects at most one information set for every player, the actions and conjectures specified by x for a subgame are unambiguous.

13. In Nöldeke and Samuelson [168], we incorrectly claimed to have shown that recurrent outcomes were simply subgame-perfect, or equivalently subgame-consistent Nash equilibria. We are grateful to Pierpaolo Battigalli for bringing this to our attention.

then cannot lead back to a state with an outcome σ^* because player ℓ prefers entering the subgame to playing her part of the outcome σ^*. Hence the learning process cannot lead back to R, which cannot then be recurrent. This establishes that recurrent outcomes must be subgame-consistent self-confirming equilibria.

Turning now to forward induction, the following proposition shows that some self-confirming equilibrium outcomes can be subgame-consistent but still fail to be recurrent because they lack a forward induction property. This failure occurs because the states supporting the subgame-consistent equilibrium outcome again allow a variety of behavior off the equilibrium path, including behavior that, once revealed, will tempt agents away from the equilibrium in quest of a higher payoff. The ability to pull agents away from an equilibrium requires only that an agent can cause a subgame to be reached that has at least one self-confirming equilibrium promising the player a higher payoff than the subgame-consistent self-confirming equilibrium. It may be impossible to support this higher payoff as an equilibrium in the entire game, so that subsequent adjustments may lead away from this payoff, but these adjustments cannot lead back to the original equilibrium. The proof of proposition 8.3 establishes the following:

Proposition 8.4 Let G be an extensive-form game with each player moving at most once along each path. Suppose outcome σ^* is recurrent. If, given a state yielding outcome σ^*, player ℓ can force entry into a subgame $G(a_\ell)$, then no self-confirming equilibrium of $G(a_\ell)$ can give player ℓ a higher payoff than does σ^*.

The result in proposition 8.4 associates a strong forward induction property with locally stable outcomes. For example, this property is stronger than that of van Damme [239], which rejects an equilibrium if player ℓ can force entry into a subgame with a *unique* equilibrium that gives ℓ a higher payoff than the original equilibrium, but that (unlike proposition 8.4) makes no comment about the case in which there are multiple such equilibria in the subgame. In many games, the conditions of proposition 8.4 will be too stringent to be satisfied by any of the subgame-consistent self-confirming equilibrium outcomes. A recurrent outcome fails to exist in such a case.

In two-player games (with each player moving at most once along any path), these results can be sharpened. We can think of such a game

	L_1	R_1
T_1	4, 1	−4, 0
B_1	0, 0	2, 1

	L_2	R_2
T_2	3, 3	−5, −5
B_2	−5, −5	1, 1

Figure 8.1
Forward induction example

G as follows: proceeding first, player I chooses which of player II's information sets h in the game G to reach, and then, a normal-form game is played with strategy sets given by player II's behavior strategies (in G) at h and by player I's strategies (in G) that cause h to be reached. We will refer to this latter representation of the game as the "extended form" of G. Notice that player I may move twice along some paths in the extended-form representation of a game, but it is a straightforward variation on proposition 8.3 that any recurrent outcome of G must correspond to a subgame-consistent, self-confirming equilibrium outcome of the extended form of G.

For generic games of this class, proposition 8.4 implies that a recurrent outcome of G corresponds to a subgame-consistent self-confirming equilibrium outcome of the extended form satisfying the never-weak-best-response (NWBR) property.[14] This follows from the fact that in such games, a subgame-consistent self-confirming equilibrium outcome σ^* fails the NWBR property only if there is an unreached subgame with a Nash equilibrium giving player I a higher payoff than σ^*. The converse does not hold, however, in that subgame-consistent self-confirming equilibria satisfying NWBR need not be recurrent.

The following example illustrates the forward induction properties of recurrent outcomes in two-player games and verifies the final claim in the previous paragraph.

Example 8.1 Consider the game whose extended form is represented in figure 8.1. In this extended form, first player I chooses one of the matrices, and then players I and II play the game represented by that matrix, with player I choosing rows and player II choosing columns. There are no dominated strategies in this game, so that

---

14. A subgame-consistent self-confirming equilibrium outcome σ^* of the extended form satisfies NWBR if it is also a subgame-perfect equilibrium outcome of the extended form obtained by removing strategies that fail to be best responses to any elements of the Nash component containing σ^*. See Kohlberg and Mertens [135].

iterated weak dominance fails to eliminate any strategies. NWBR eliminates the subgame-perfect equilibrium (T_2, L_2). In particular, B_1 is a best response to no element in the component supporting outcome (T_2, L_2). Once B_1 is eliminated, subgame consistency requires player II to choose L_1 at her first information set, at which point (T_2, L_2) ceases to be an equilibrium. Notice that NWBR cannot eliminate the subgame-perfect equilibrium given by (B_2, R_2) because every strategy is a best response to some element in the component supporting this outcome.

In contrast, the only recurrent outcome in this game is (T_1, L_1). To see this, consider a singleton absorbing set x supporting either outcome (B_2, R_2) or (T_2, L_2). By proposition 8.2, any recurrent set containing x contains a state y in which all player-II agents choose L_1 at their first information set. Now let a mutation occur that causes a player-I agent to play T_1. Then with positive probability, all player-I agents learn and switch to T_1, yielding a new singleton absorbing set x' that supports outcome (T_1, L_1). This establishes $M(y) \cap B(x') \neq \emptyset$ (cf. definition 7.4). However, the equilibrium (T_1, L_1) gives player I the highest possible payoff in the game, and action L_1 is a strict best reply for player II at the relevant information set. There is then a set R, containing x', that contains only states yielding the outcome (T_1, L_1) and that is recurrent for all sufficiently large population sizes. This implies that the state x supporting outcome (B_2, R_2) is not contained in a recurrent set. Furthermore, (T_1, L_1) is the *unique* recurrent outcome because an argument similar to the one just given shows that neither the outcome (B_1, R_1) nor a mixed-strategy equilibrium outcome is recurrent.   □

8.4   Backward Induction

Even if a game admits a unique subgame-consistent self-confirming equilibrium outcome σ^*, the results of the previous section do not allow us to conclude that σ^* appears in the limit distribution, much less that σ^* is the only outcome in that distribution. The only possible recurrent outcome is σ^*, but a recurrent outcome may fail to exist, or the corresponding recurrent set may not constitute the entire support of the limit distribution. To address these issues, we shall consider extensive-form games with each player moving at most once along each path, as before, but now restrict attention to generic games of perfect information. The result of this section is that the subgame-perfect equilibrium must appear in the limiting distribution, but only under stringent conditions will it be the only outcome in the limiting distribution.[15]

The genericity condition is that for every player $\ell \in I$ and every payoff of distinct terminal nodes ξ and ξ', $\pi_\ell(\xi) \neq \pi_\ell(\xi')$. It is well known that such games have a unique subgame-perfect equilibrium.[16]

A useful preliminary result is that such games have only singleton absorbing sets, which in turn implies that the evolutionary process converges to a recurrent set containing only states yielding self-confirming equilibrium outcomes:[17]

Proposition 8.5 Let G be a generic extensive form game of perfect information with each player moving at most once along each path. Then all absorbing sets are singletons.

(This and all subsequent propositions in this chapter are proved in section 8.7.)

Next, we can show that there is a unique recurrent set, and that recurrent set contains the subgame-perfect equilibrium outcome. From proposition 7.7, this implies that the limit distribution satisfies a backward induction property in the sense that it must assign strictly positive probability to the subgame perfect equilibrium. Furthermore, if the subgame-perfect equilibrium outcome is recurrent, then it must be the unique outcome appearing in the support of the limiting distribution. The proof of this result, which combines ideas from the proof of propositions 7.7 and 8.3, exploits the fact that absorbing sets are singletons.

Proposition 8.6 Let G be a generic extensive-form game of perfect information with each player moving at most once along each path.

---

15. An equilibrium is subgame-perfect in such a game if and only if it is a subgame-consistent self-confirming equilibrium, and we can accordingly speak simply of subgame-perfect equilibria. These games are dominance-solvable, so that the subgame-perfect equilibrium is the outcome of the iterated elimination of weakly dominated strategies in the normal form. The finding that the subgame-perfect equilibrium will often not be the only outcome in the limiting distribution is then related to the finding in chapter 7 that a similar model need not eliminate weakly dominated strategies in the normal form.

16. Notice that the unique subgame-perfect equilibrium will always be in pure strategies, so that the question of whether mixed strategies can be duplicated by population proportions does not arise.

17. Canning [59] shows that all the absorbing sets for a fictitious play process in extensive-form games of perfect information are singletons. In a model with "trembling hand" trembles of the type introduced by Selten [208], but without mutations, Canning finds the subgame-perfect equilibrium is the unique limiting outcome (as trembles become small).

Then there is a unique recurrent set and that recurrent set contains the subgame-perfect equilibrium outcome.

Proposition 8.6 establishes that the subgame-perfect equilibrium is the only candidate for a recurrent outcome. When will it be recurrent? Proposition 8.4 showed that the subgame-perfect equilibrium outcome will *not* be recurrent if some player can be tempted away from the equilibrium path by a self-confirming equilibrium in a subgame that promises that player a higher payoff. This suggests that if no such opportunities arise, so that no player can force entry into a subgame with at least one self-confirming equilibrium offering the player a higher payoff than the subgame-perfect equilibrium, then the latter will be recurrent. The following example shows that this conjecture is false.

Example 8.2   Consider the three-player game G^* shown in figure 8.2.[18] The unique subgame-perfect equilibrium is given by the strategy combination (R, R, R), for a payoff of $(1, 1, 1)$. Let x_R be the state in which all agents from all populations play R, and let x_L be a state in which all agents play L, with conjectures matching these actions. Note that x_L is a self-confirming equilibrium, and gives payoffs $(0, 0, 0)$. We can now show that x_L must be contained in the support of the limit distribution.

By proposition 8.6, x_R is in the support of the limit distribution, and so (by proposition 8.2) is the state in which all agents from populations *I* and *II* play *R* and all agents from population *III* play *L*, with conjectures matching the behavior of agents at reached decision nodes and with agents from population *I* and *II* conjecturing that all agents from population *III* play *R*. Now let a single mutation cause an agent from population *II* to play *L*. With positive probability all agents from population *II* (but no other agents) now receive the learn draw, update their conjectures to match the observation that all agents from population *III* play *L*, and switch to playing a best reply of *L*. Suppose in the next stage of the evolutionary process all agents from population *I* (but no other agents) learn. They update their conjectures to match the observed behavior of other agents, which is that agents from population *II* are playing L, and then switch to their best reply of playing *L*.

---

18. It is easy to establish that for all two-player games (from the class of games under consideration) the subgame-perfect equilibrium outcome is recurrent for every sufficiently large population size, and is thus the only outcome in the limiting distribution of the evolutionary process.

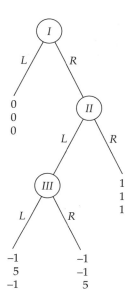

Figure 8.2
Game G^*

It is easily checked that the resulting state is a singleton absorbing set yielding the same outcome as x_L. Beginning with a state contained in the recurrent set containing x_R, a single mutation thus reaches the basin of attraction of a state yielding the outcome corresponding to x_L. From the fact that x_R is contained in a recurrent set and from proposition 8.2 again, it thus follows that x_L and x_R are contained in the same recurrent set, and consequently the self-confirming equilibrium x_L must appear in the support of the limit distribution.   □

This example illustrates the differences between the outcomes of this model and various notions of evolutionary stability. It will be helpful to illustrate these differences by considering Swinkels's equilibrium evolutionarily stable (EES) sets [230]. It is easy to show that in game G^*, the (unique) EES set and the limiting distribution of the current model both contain a set of strategies corresponding to the payoff $(1, 1, 1)$. In our case, the recurrent set containing these states also contains states corresponding to any possible mixture on the part of player III, including mixtures that violate the conditions for a Nash equilibrium. It is the presence of these states that prevents the subgame-perfect equilibrium from being a recurrent outcome because single mutations can lead from these states to learning dynamics that draw the system away from the subgame-perfect outcome. The corresponding EES set allows only Nash equilibria, and hence includes mixtures on the part of player III only if a probability of $\frac{1}{3}$ or less is placed on L. The dynamic intuition behind the EES set is that deviations from Nash equilibrium, even at unreached subgames, prompt immediate responses that push the system back to the EES set, so that the system is not free to drift away from Nash equilibria via movements at unreached information sets.

Swinkels [231] examines a dynamic model in which he obtains results similar to those of the current chapter in a much different context. Swinkels considers an asymptotic stability notion for sets of outcomes under a dynamic process specified by a deterministic differential equation, and shows that for a broad class of deterministic dynamics, sets satisfying this assumption must contain a Kohlberg and Mertens stable and hyperstable set [135]. The current model thus shares with Swinkels the characteristic that "good" outcomes, such as subgame-perfect or stable equilibria, are contained in the limiting outcome, but the limit outcome will generally not be restricted to include only such good outcomes. In the current model, if any self-confirming equilibrium strategy profile appears in the limiting distribution, then so do any other strategy profiles created by altering behavior at unreached subgames. These can include strategy profiles that are not self-confirming equilibria but are surrounded by learning dynamics that lead to quite different outcomes, causing the latter to also appear in the limiting distribution. In Swinkels's dynamic model, similar forces appear. Every equilibrium appearing in the outcome is accompanied by a host of strategy profiles that give the same outcome, in some cases including profiles surrounded by dynamics that lead to other outcomes. Swinkels observes that his model holds some hope of yielding asymptotically stable sets that contain only subgame-perfect outcomes if the dynamics have rest points only at *Nash* equilibria (as opposed, for example, to having rest points at all self-confirming equilibria, as in the current model).[19]

Proposition 8.4 and example 2 suggest that the subgame-perfect equilibrium outcome will be the unique outcome in the limit distribution of the evolutionary process if no player can be "tempted" away from the subgame-perfect equilibrium by a higher payoff, even

---

19. The potential difficulty with such an assumption is that it seemingly requires players to react to changes in actions at information sets that are not reached during play.

a nonequilibrium payoff. To make this intuition precise, it is helpful to have a bit more terminology. The subgame-perfect equilibrium outcome, identified by the terminal node ξ^*, is a *strict outcome* (cf. Balkenborg [4]) if no player can force entry into a subgame that has a terminal node with a higher payoff than the subgame-perfect equilibrium. More formally, suppose that, given the subgame-perfect equilibrium outcome, player ℓ can force entry into the subgame $G(a_\ell)$. Let $\Xi(a_\ell)$ denote the set of terminal nodes contained in $G(a_\ell)$. If ξ^* is a strict outcome, then for all $\xi \in \Xi(a_\ell)$,

$$\pi_\ell(\xi^*) > \pi_\ell(\xi)$$

must hold. And we have[20]

Proposition 8.7 Let the game be as in proposition 8.6. If the subgame-perfect equilibrium outcome is strict, it is the unique outcome in the limiting distribution for all sufficiently large population sizes.

These results provide both good and bad news for backward induction. Subgame perfection presents the only possibility for a recurrent outcome, and the subgame-perfect equilibrium always appears in the limit distribution of our simple games. In many games, however, the limit distribution contains states that do not yield the subgame-perfect outcome path. This suggests a reinterpretation of the role of backward induction in finite extensive-form games of perfect information, with subgame perfection being a sufficient but not necessary condition for an outcome to be interesting.

8.5   Forward Induction

In the spirit of the previous section, this section examines a class of games simple enough to give us some insight into the question of when the forward induction properties of recurrent outcomes carry over to the limiting distribution of the evolutionary process. Our attention will be restricted to two-player games whose extended form first calls for player I to choose between an outside option, yielding payoff vector π^*, and entering a normal-form game $\tilde{G}$.

---

20. Notice that propositions 8.4 and 8.7 are not converses of one another because proposition 8.7 requires that no terminal node in $G(a_\ell)$ give player ℓ a higher payoff than $\pi_\ell(\xi^*)$, while proposition 8.4 requires that a self-confirming equilibrium outcome in $G(a_\ell)$ give a higher payoff than $\pi_\ell(\xi^*)$.

The analysis of these games is complicated by the fact that they may have nonsingleton absorbing sets. Considerable progress can still be made, but the following assumptions strengthen the results.

Assumption 8.1.1   If an agent α from population ℓ receives the learn draw, and (after updating conjectures) finds that she is not playing a best reply, and if other player ℓ agents are playing best replies, then agent α switches to a best reply played by one of the other player ℓ agents.

Assumption 8.1.2   No Nash equilibrium of game $\tilde{G}$ gives player I the same payoff as the outside option.

Assumption 8.1.1 introduces some further inertia into the learning process, indicating that when switching strategies, agents will choose best replies that are already played by other members of their population if such strategies exist. This assumption is likely to be reasonable in contexts where learning is at least partly driven by imitation. Assumption 8.1.2 is a mild genericity assumption.

Proposition 8.8   Suppose that there exists a *strict* Nash equilibrium of the game $\tilde{G}$ providing player I with a payoff higher than π_I^*.

(i) Then for all sufficiently large population sizes, the limiting distribution does not contain a state in which all agents in population I choose the outside option.

(ii) If assumptions 8.1.1–8.1.2 hold, then for all sufficiently large population sizes, the limiting distribution does not contain a state in which *any* agent from population I chooses the outside option.

The intuition behind this result will by now be familiar. Consider a state where all agents in population I take the outside option. If a recurrent set contained this state, it would also contain a state where the outside option is taken, but where play in game $\hat{G}$ corresponds to the Nash equilibrium of the game that gives player I a higher payoff than π_I^*. A mutation causing an agent from population I to enter game $\hat{G}$, followed by learning, will then cause the outside option to be abandoned.

Proposition 8.8 can be contrasted with the forward induction concept introduced by van Damme [238, 239]. Van Damme's notion stipulates that if there is one and only one Nash equilibrium of the subgame that offers player I a higher payoff than the outside option (with other Nash equilibria of the subgame offering lower payoffs), then equilibria in which player I chooses the outside option will not appear.

The result established in proposition 8.8 differs in two respects. First, it does not require that the equilibrium in the game $\tilde{G}$ providing i with a higher payoff than adhering to the outside option be unique.[21] The uniqueness requirement appears in van Damme's definition because of a presumption that if there are multiple equilibria in the subgame that dominate the outside option for player I, then entering the subgame provides an ambiguous signal and the forward induction argument loses its force. Ambiguity is not a difficulty in the evolutionary approach. The evolutionary process "assigns" a meaning to the event of entering the subgame, even if this meaning is a priori ambiguous.

Second, although the existence of a Nash equilibrium in the subgame exhibiting certain properties appears as a condition of the proposition, the result does not require an assumption of equilibrium play in the subgame, and hence requires no assumption of equilibrium play after nonequilibrium actions. In particular, even though the result establishes that the outside option cannot appear in the limiting distribution, it does *not* follow that the limiting distribution must correspond to an equilibrium of $\tilde{G}$. Instead, the possibility cannot be excluded that the limit distribution corresponds to a recurrent set containing some non-singleton absorbing set (that contains only states in which player I enters $\tilde{G}$ and receives a higher payoff than the outside option).

One of the apparent successes of evolutionary game theory has been the use of either refinements of the evolutionarily stable strategy concept or dynamic models to establish forward induction properties for outcomes of cheap talk games. Bhaskar [15], Blume, Kim, and Sobel [34], Kim and Sobel [133], Matsui [148], Sobel [223], and Wärneryd [246] establish conditions under which the evolutionary process, operating on the cheap talk game, selects efficient equilibria of the underlying game. The forces behind these results are quite similar to those behind proposition 8.8.

To see this similarity, suppose that the underlying game is the Stag-Hunt Game of figure 2.9. This game has two Nash equilibria, $(X, X,)$ and (Y, Y), with the former being payoff-dominant and the latter risk-dominant. Now suppose that before playing the game, each player has

---

21. Our requirement that this equilibrium be strict plays a role similar to van Damme's assumption that the game is generic [238, 239].

an opportunity to announce either "X" or "Y," with the announcements being made simultaneously. These can be interpreted as announcements of strategies that the agents claim they will play, but the announcements are "cheap talk" in the sense that they impose no restriction on the action that the player actually takes. A strategy is now an announcement and a specification of what the player will do for each possible announcement configuration. Hence the strategy "XYX" is interpreted as "Announce X, play Y if the opponent announces X, and play X if the opponent announces Y." Let this game be played by a single population of players who are randomly chosen to be row or column players when matched.

Consider an outcome in which everyone sends the message Y and then plays the inefficient equilibrium (Y, Y). This is the counterpart of all agents choosing the outside option in proposition 8.8. Now let mutations switch all agents to strategies in which message Y is sent and action Y played, but specifying that the strategy X would be played in any match where at least one agent sent the message X. These mutations do not affect realized play, do not give rise to any learning pressure on the system, and do not take the system outside the mutation-connected component of absorbing states containing the original state in which all agents send Y and play Y. Now let a mutation introduce a strategy in which an agent sends the currently unused message X. The resulting dynamics will lead to an outcome with only the efficient equilibrium being played. As in the case of proposition 8.8, this argument has proceeded by noting that the recurrent set supporting the inefficient equilibrium includes states in which an unreached subgame exhibits equilibrium behavior yielding higher payoffs than current play, and then noting that a mutation, leading to that subgame, causes agents to adopt this behavior.

The key step in making this argument for the cheap talk case is ensuring that an unused message exists. This is especially problematic if agents are allowed to use mixed strategies, where they can attach positive probability to all messages while playing the same inefficient equilibrium after any message pair, effectively babbling away the opportunity to coordinate. An important ingredient in evolutionary analyses of cheap talk is then typically some variant of a "large message space" assumption ensuring that unused messages exist.

The sequence of mutations required to produce an efficient outcome in a game of cheap talk, or to produce the forward induction outcome in proposition 8.8, may be quite special. The length of time required for such a sequence to appear may accordingly be very long; these are then clearly ultralong-run results. Chapter 6 shows that forward induction arguments may fare quite differently in the long run.

8.6   Markets

Backward and forward induction are the two basic building blocks of equilibrium refinements. The results of this chapter's ultralong-run evolutionary model offer mixed support for backward induction. We have examined the class of games in which backward induction is most likely to work, namely, finite extensive-form games of perfect information in which each player moves at most once along any equilibrium path, using a best-reply-based learning model that is also conducive to backward induction. Although the limit distribution of the resulting evolutionary process will include the unique subgame-perfect equilibrium of such a game—the only possibility for a unique limiting outcome—in the absence of strong conditions, it will also include self-confirming equilibria that are not subgame-perfect.

Self-confirming equilibria that are not subgame-perfect appear in the limiting distribution because the evolutionary system can be tempted away from the subgame-perfect equilibrium outcome by the lure of higher payoffs. In games with multiple subgame-perfect equilibria, these same forces, by allowing the system to be drawn away from some of the equilibria, can produce forward induction properties. This suggests that in an evolutionary analysis, backward induction is likely to play a role somewhat different from its usual one, being sufficient but not necessary to identify an outcome as interesting, while forward induction is likely to play a larger than usual role.

This characterization is based on a quite specific model, and a wide variety of other models must be investigated before general conclusions can be drawn. An immediate challenge facing such an investigation is the examination of more complicated games. The games we have examined in this chapter are simple enough that the ability of players to observe only outcomes and not strategies has not caused great difficulties, but the issues become more complicated in larger games.

The key to constructing useful models of more complicated games may involve more detailed modeling of the process by which agents interact. An integral feature of an evolutionary model is the possibility that players may learn both from their own experience and the experience of others. Our evolutionary model, like most, does not discuss the details of how players acquire information about the outcomes of games played by others.[22] It is assumed that players are randomly matched in groups, with one representative from each population, to play games, and is then further assumed that players who learn automatically observe or know the outcomes of all matches.

In practice, the mechanism by which agents are brought together to play a game may bear little resemblence to a random matching process that arranges agents into temporary, isolated groups, with each such group being equally likely and each possible group forming in each period. The alternative arrangements by which the game is played may play an important role in determining how information is transferred between agents. For example, it may often be more useful to think of agents as interacting in markets rather than in pairs. Information may then be conveyed by the market outcome, such as the market price.

Nöldeke and Samuelson [169] examine an evolutionary model of play in signaling games. Rather than assuming that players meet in pairs or triplets to play the game, the model calls for all players to participate in a single market in each period. A key aspect of the information transmission process is then the senders' observation of market prices, which not only imposes some useful structure on the evolutionary process but, in this model, also produces a dynamic model virtually identical to that outlined by Spence [225, 226] in his introduction of signaling. As a result, the forces driving the equilibrium selection results in the evolutionary analysis can be directly linked to ideas appearing in Spence's analysis and the subsequent equilibrium refinements literature. I suspect great progress can be made in applying evolutionary models through additional work on modeling the market mechanisms by which agents are matched to play games and by which information is transferred.

---

22. The assumption that all agents play all other agents in each period does not suffice to provide this information. Some of agent i's actions may be revealed only when agent i plays agent j, ensuring that agent k cannot learn everything there is to be learned about agent i simply from observing the outcomes reached when agent k plays agent i.

8.7   Appendix: Proofs

Proof of Proposition 8.2
Part (i)   Let x and x' have the specified property. Then x and x' feature identical conjectures at all information sets and identical actions at every information set reached during the play of any match. Because all of the actions taken in state x are best replies to agents' conjectures, the same is true of x', ensuring that x' is a singleton absorbing set. To show that x' lies in $R(x)$, the recurrent set containing x, we need only note that we can construct a sequence of singleton absorbing sets beginning with x and leading to x', each successive pair of which is adjacent; by altering, one agent at a time, the actions at unreached information sets that prevail in state x to match those of state x'.   □

Part (ii)   A similar argument establishes that x' in this case is contained in $R(x)$ because altering the conjectures of agents from populations other than ℓ, in a subgame that they cannot cause to be reached, does not affect the optimality of their actions at reached information sets.   □

Proof of Proposition 8.3   Let σ^* be a recurrent outcome generated by the recurrent set R consisting of singleton absorbing sets. From proposition 8.1, the outcome σ^* must then be the outcome of a self-confirming equilibrium. Suppose σ^* does not correspond to a subgame-consistent self-confirming equilibrium. Then every state $x \in R$ satisfies the condition that there exists a player ℓ who can force entry into a subgame $G(a_\ell)$ such that every self-confirming equilibrium of $G(a_\ell)$ yields ℓ a higher payoff than does σ^*. Proposition 8.2 ensures that R contains a state z in which the actions and conjectures of all players who have an information set in $G(a_\ell)$ correspond to a self-confirming equilibrium of $G(a_\ell)$.[23] Now consider a mutation in the actions of a player ℓ agent that causes $G(a_\ell)$ to be reached. Then with positive probability, all player ℓ agents learn. These agents will switch their actions so as to enter $G(a_\ell)$. The learning mechanism then cannot further adjust actions or conjectures in $G(a_\ell)$ because these constitute a self–confirming equilibrium on $G(a_\ell)$. This in turn ensures that the

---

23. This step uses the assumption that each player moves at most once along every path, which ensures that player ℓ has no information set in $G(a_\ell)$. Proposition 8.2 then implies that we can choose the desired conjectures for all players having an information set in $G(a_\ell)$.

learning mechanism leads to an absorbing set, say A, that contains at least one state not resulting in the outcome σ^*. We then have $A \notin R$ and $M(z) \cap B(A) \neq \emptyset$, contradicting the claim that R is recurrent (cf. definition 7.4). $\square$

Proof of Proposition 8.5 Let x be a state that is not a self-confirming equilibrium. It suffices to show that with positive probability, the learning mechanism leads from x to a state that yields a self-confirming equilibrium (and is hence a singleton absorbing set). This implies that every state that is not a self-confirming equilibrium lies in the basin of attraction of a self-confirming equilibrium, and thus the only absorbing sets are singletons consisting of self-confirming equilibria. To show this, fix x and consider the evolutionary sequence in which, in each period, all agents receive the learn draw. Let $X_1, \ldots, X_m$ be a sequence of sets of nodes constructed by letting X_1 consist of all nodes that are followed only by terminal nodes, and letting X_i for $i > 1$ consist of all nodes not contained in $X_1, \ldots, X_{i-1}$ and followed only by nodes contained in $X_1, \ldots, X_{i-1}$ or terminal nodes. Let t_1 be large enough that if x_1 is a node of X_1 and if there are at least two periods in which some match reaches node x_1, then x_1 has already been reached in at least two periods prior to t_1 (since X_1 is finite and the sequence of learn draws has been fixed, t_1 exists). Then no conjectures about actions at any node $x_1 \in X_1$ can change after period t_1: either node x_1 has already been twice reached, in which case actions (after the first time the node is reached) and conjectures (after the second time) must match the unique optimal action at this node and cannot subsequently be changed (because the actions in question can never be suboptimal);[24] or the node is never again reached, in which case conjectures cannot be subsequently changed by the learning process. Now let X_2 be the collection of nodes followed only by either terminal nodes or nodes in X_1. A similar argument shows that there exists a time t_2 after which conjectures at nodes in X_2 cannot change. Continuing this argument establishes the existence of a (finite) period t_m after which no conjectures can change. In period $t_m + 1$, all agents will switch to best responses to these conjectures, with these best responses confirming the conjectures (because there is no further conjecture adjustment), yielding a

---

24. This step uses the assumption that each player moves only once along each path, so that if player ℓ moves at this node, then conjectures about player ℓ's behavior at this node cannot subsequently be altered by having player ℓ change an action at a prior node that previously made this node unreachable.

self-confirming equilibrium. Hence the learning process leads to a self-confirming equilibrium with strictly positive probability, establishing the desired result. □

Proof of Proposition 8.6 Let R be a recurrent set. It is first shown that R must contain a state yielding the subgame-perfect equilibrium outcome. A simple extension of the argument establishing proposition 8.2 then implies that R must contain all singleton absorbing states corresponding to the subgame-perfect equilibrium outcome, thus ensuring that no other recurrent set exists and yielding the result. To show that R must contain a state yielding the subgame-perfect equilibrium outcome, consider an arbitrary singleton absorbing set x contained in R. Let $X_1, \cdots, X_m$ be the sequence of nodes constructed in the proof of proposition 8.5. Let n_x be the smallest integer such that state x prescribes play at a node θ_{n_x}, reached in the course of play, that is contained in the set X_{n_x} but does not match play prescribed by the subgame-perfect equilibrium. (If there is no such n_x, then x yields the subgame-perfect equilibrium outcome and we are done). By proposition 8.2, R contains an element, say x', with $n_{x'} = n_x$ and with x and x' yielding identical play and conjectures and with x' prescribing actions corresponding to the subgame-perfect equilibrium at every node that is not reached. Consider x'. Let a single mutation occur that causes an agent at node θ_{n_x} to switch to the subgame-perfect equilibrium action. Then with positive probability, all agents who move at node θ_{n_x} learn and switch to the subgame-perfect equilibrium action at node θ_{n_x}. Any subsequent learning, because it cannot alter actions following node θ_{n_x} (or any other nodes followed by subgame-perfect equilibrium actions) will then lead to a state x_1, which is either the subgame-perfect equilibrium or satsifies $n_{x_1} > n_x$. (This inequality follows from noting that x must correspond to a pure strategy, given our genericity assumption and that x is self-confirming; thus all nodes in X_{n_x} other than θ_{n_x} are unreached.) Because R is recurrent, we have x_1 in R. As this argument can be applied to any $x \in R$ and the set of nodes (and hence m) is finite, this ensures that the subgame-perfect equilibrium is in R. □

Proof of Proposition 8.7 Let $\bar{\pi}_\ell$ be the maximal payoff that could result for player ℓ if he deviates from the subgame-perfect equilibrium path. Let ξ^* be the subgame-perfect equilibrium outcome. Because the subgame-perfect equilibrium outcome is a strict outcome, we have $\bar{\pi}_\ell < \pi_\ell(\xi^*)$. Let $\underline{\pi}_\ell = \min_{\xi \in \Xi} \pi_\ell(\xi)$, that is, $\underline{\pi}_\ell$ is the worst possible payoff player ℓ could receive. Define $\bar{N}_\ell = \frac{\pi_\ell(\xi^*)-\pi_\ell}{\pi_\ell(\xi^*)-\bar{\pi}_\ell}$ and let $\bar{N} = \max_\ell \bar{N}_\ell$. Consider a population size $N > \bar{N}$ and a singleton absorbing set x^* yielding the subgame-perfect equilibrium outcome. By propositions 8.5 and 8.6 it suffices to show that, given x^*, a single mutation cannot yield a state lying in the basin of attraction of a self-confirming equilibrium that does not yield the subgame-perfect equilibrium outcome. This ensures that the subgame-perfect equilibrium outcome is the only outcome in any recurrent set, yielding the result. Suppose that a single mutation occurs and call the resulting state x'. If this mutation changes the characteristic of an agent of player ℓ who cannot force entry into a subgame, then x' yields the subgame-perfect equilibrium outcome. Hence we may assume that the mutation changes the characteristic of an agent, whom we refer to as the "affected agent," of player ℓ who could force entry into a subgame $G(a_\ell)$. Suppose this mutation does not cause the affected agent to change his strategy on the equilibrium path. Then the current state remains unchanged until the affected agent receives the learn draw, and if he does so, he will update his conjecture to match the observed behavior of other agents. Because the subgame-perfect equilibrium outcome is a strict outcome, the resulting state must be a self-confirming equilibrium yielding the subgame-perfect equilibrium path. Finally, suppose the mutation causes the affected agent to force entry into a subgame $G(a_\ell)$. By construction, it is the case that for any $N > \bar{N}$, all agents but the affected agent will, upon receiving the learn draw, not change their action. Hence the only adjustment in actions that can be caused by the learning process is that the affected agent eventually switches back to the subgame-perfect equilibrium path, completing the proof that the subgame-perfect equilibrium outcome is the only outcome in a recurrent set.   □

Proof of Proposition 8.8

Statement (i)   Let x' be a singleton absorbing set corresponding to the Nash equilibrium of the game $\tilde{G}$ that yields player I a higher payoff than π_I^*. Because this Nash equilibrium is assumed to be strict, $\{x'\}$ is a recurrent set for sufficiently large N. To show that the limiting distribution excludes states in which all player I agents play the outside option, it suffices to show that if an absorbing set A containing such a state y is contained in a recurrent set $R(y)$, then $M(R(y)) \cap B(x') = 1$, a contradiction. Thus consider a state y in which all player I agents play the outside option. If y is a singleton absorbing set, then it follows by a simple variation of the argument in proposition 8.6 that

$M(R(y)) \cap B(x') = 1$. (In particular, $R(y)$ contains a state in which all player II agents play their part of the equilibrium of $\tilde{G}$ whose player-I payoff exceeds π_I^*; a single mutation causing a player-I agent to enter $\tilde{G}$ then yields a state in the basin of attraction of x'.) Hence a singleton absorbing set in which all agents from population I choose the outside option cannot appear in the limit distribution. To complete the proof, it suffices to show that a state y in which all player-I agents choose the outside option cannot be part of a nonsingleton absorbing set. Suppose that it is. Then this absorbing set, denoted A, must contain a state y' in which not all player-I agents play the outside option (because otherwise A would be a singleton), with $F_{y'y} > 0$ (because the learning mechanism leads with positive probability from any state in a nonsingleton absorbing set to any other). $F_{y'y} > 0$ can hold only if the outside option is a strict best reply in state y' or if it is a weak best reply and no player-I agents currently play best replies. Let either of these conditions hold. Then there is a positive probability that all (and only) player-I agents learn in state y', with these agents switching to the outside option. This yields a singleton absorbing set (because the outside option is by hypothesis a best reply for player-I agents and any strategies for player-II agents are best replies to the outside option). Because y is a singleton absorbing set, it cannot be that A is a nonsingleton absorbing set because $y \in A$, yielding a contradiction.

Statement (ii) Let assumptions 8.1.1–8.1.2 hold. Then we show that there cannot exist a nonsingleton absorbing set containing a state in which some (though possibly not all) player-I agents play the outside option. Suppose such a set exists, denoted A. Then A must contain two states, say x and x', such that no player-I agent plays the outside option in state x', some player-I agents play the outside option in state x, and $F_{x'x} > 0$.[25] Then the outside option must be a best reply for player-I agents in state x'. If it is a unique best reply (or if it is a weak best reply, but no other best reply is currently played), then with positive probability all (and only) player-I agents learn and the learning mechanism

---

25. If not, then the outside option must be a best reply in every state in A. However, there cannot exist a set of strategies for a player that are best replies in every state of a nonsingleton absorbing set. If such a set S existed, then there must exist a state in the absorbing set A where all player-I agents play strategies in S. However, no subsequent adjustments in player-I strategies can occur because the strategies in S are best replies in every state in A. With positive probability, the other player's agents then all learn and switch to best replies, yielding a self-confirming equilibrium and hence a singleton absorbing set, contradicting the fact that A is an absorbing set.

leads to a self-confirming equilibrium and hence singleton absorbing set in which all player-I agents choose the outside option, contradicting the assumption that A is a nonsingleton absorbing set. If some other best reply is currently played, then under assumption 8.1.1 no player-I agents can switch to the outside option, contradicting $F_{x'x} > 0$. Hence no nonsingleton absorbing set can contain states in which some player-I agents play the outside option. The proof is then completed by noting that, given assumption 8.1.2, there cannot exist a *singleton* absorbing set in which some but not all player-I agents play the outside option.   □

Chapters 7–8 have shown that even small amounts of noise, in the form of mutations that randomly switch agents between strategies, can have important implications for the stationary distributions of evolutionary models. This echoes the findings of Kandori, Mailath, and Rob [128] and of Young [250], who show that accounting for the effects of improbable mutations leads to the prediction that the limit distribution concentrates all of its probability on the risk-dominant equilibrium in 2×2 symmetric games.

These results raise two questions. (1) What happens if we incorporate other sources of noise, besides arbitrarily improbable mutations, into the model? (2) How long is the ultralong run to which these limit distribution calculations can be applied?

To motivate question 1, notice that like all models, an evolutionary model is an approximation of complicated phenomena, ideally including salient features of the underlying process and excluding unimportant ones. These unimportant, excluded features will cause the performance of the underlying process to deviate from that of the model in a variety of ways, which is to say, the model excludes various sorts of noise. One hopes that this uncaptured noise is sufficiently small relative to the forces included in the model as to have only a minor effect on outcomes. Mutation-counting models, however, examine the limit as the probability of a mutation goes to zero. As this source of noise becomes small, might it not be overwhelmed by other factors excluded from the model? Can this cause the process to yield results that differ from the predictions of a model based solely on mutations?

The relevance of question 2 is straightforward. In order for the limit distribution to be an applicable description of the state of the process, we must wait long enough for mutations to induce transitions between states that cannot occur under the learning process alone. But if the probability of a mutation is very small, this may entail waiting a very long time.

In this chapter we shall introduce another source of noise into the model.[1] The point of departure for the analysis is the observation that in the models of Young [250] and of Kandori, Mailath, and Rob [128] as well as in our model of chapter 8, agents choose best responses given their information whenever the selection mechanism (but not mutations) prompts them to choose strategies. I shall describe such agents as "maximizers."

We shall consider a selection mechanism that is motivated by a simple belief: people make mistakes. It may be that people are more likely to switch to a best reply than otherwise, but they are unlikely to be so flawless that they *always* switch to a best reply when reassessing their strategies. Furthermore, I do not expect these mistakes to be negligible, and hence do not think it appropriate to examine the limiting case as the mistakes become arbitrarily small. I refer to agents who are plagued by such mistakes as "muddlers." These mistakes might seem absurd in the stark models with which we usually work, but arise quite naturally in the noisy world where games are actually played.

The Aspiration and Imitation model, introduced in chapter 3, is a muddling model because an unlucky draw from the random variables determining realized payoffs or aspiration levels may prompt agents to switch away from a best response. Much of the analysis of chapter 7 was conducted within the context of a muddling model because there needed to be only positive probability that the selection mechanism move agents in the direction of a best response, though the agents of the previous chapter were maximizers.

This chapter examines the effect of muddling agents in 2×2 symmetric games played by a single population. Examining muddlers rather than maximizers has the consequence that the expected waiting time before the ultralong-run predictions of the model become relevant is reduced. To see why, consider the possibility that a population of agents has found its way to an equilibrium not selected in the ultralong run. In the maximizing models of Young [250] and of Kandori, Mailath, and Rob [128], a large number of *simultaneous* mutations are now necessary for the system to escape from the equilibrium's basin

---

1. The chapter reports work in progress; for a more detailed report, including proofs, see Binmore and Samuelson [24].

of attraction. In contrast, the Muddling model requires only one mutation to escape from the equilibrium, after which the agents may *muddle*
their way out of its basin of attraction.

Incorporating noisy learning into the model also has implications for
equilibrium selection: muddling models do not always select the same
equilibria as maximizing models. In the symmetric 2×2 games studied in this chapter, the maximizing models of Young [250] and of Kandori, Mailath, and Rob [128] choose between two strict Nash equilibria
by selecting the risk-dominant equilibrium. When risk dominance and
payoff dominance conflict, the Muddling model sometimes selects the
payoff-dominant equilibrium.

The world is an inherently noisy place, and the results of this chapter
again lead to the conclusion that noise matters[2] and that we must exercise some caution in performing the convenient limiting operations
that lie at the heart of many evolutionary analyses. When examining
the limit as noise disappears, attention must be devoted to the question
of whether the way has not simply been cleared for some other source
of noise, currently excluded from the model, to take its place.

As in chapter 6, the conclusion that outcomes can be quite sensitive
to different specifications of very small perturbations appears to pose a
serious challenge to evolutionary theories. What hope is there for using
these models to describe behavior? In chapter 6 the response was to
develop comparative static implications of the model that would hold
regardless of the specification of drift, as long as drift was fixed. The
final sections of this chapter develop some implications of the model
that can potentially be compared to how agents play games.

9.1   A Muddling Model

Consider the symmetric 2×2 game $\hat{G}$ of figure 3.1, reproduced in figure 9.1. Assume that there is a single population containing N agents.
As in chapters 7–8, our interest is directed to the stationary distribution
of the learning process.

This 2×2, single-population game is much simpler than the games
examined in chapter 7, although the question of interest is now the
choice between strict Nash equilibria rather than evaluating equilibria
with alternative best replies. The former question is sufficiently more

---

2. Another perspective on how noise matters in the model of Kandori, Mailath, and Rob
[128] is provided by Bergin and Lipman [13].

	X	Y
X	A,A	C,B
Y	B,C	D,D

Figure 9.1
Game $\hat{G}$

difficult, especially in a muddling model where we do not have available the convenient assumption that strategy adjustments are always in the direction of a best response, that we cannot simply apply the techniques developed in chapter 7. To gain some additional structure, this chapter examines the continuous-time limit in which time intervals are arbitrarily short and agents' strategy adjustments are independent, idiosyncratic events.

Time is divided into discrete intervals of length τ. In each time period, an agent is characterized by the strategy X or Y she is programmed to use in that period. The model will again be a Markov model, in that only the current state affects the actions of an agent who evaluates her strategy and potentially chooses a new strategy.

The model has four parameters: the time t at which the system is observed, the length τ of a time period, the population size N, and the mutation rate λ. The analysis will involve taking limits with respect to at least three of these parameters (possibly excepting λ). Chapter 3 focused on the importance of the order in which these limits are taken. In this chapter, we shall study the ultralong-run behavior of the system, represented by taking the limit $t \to \infty$. We then take the limit $\tau \to 0$. This gives a model in which agents revise their strategies at uncoordinated, idiosyncratic times. Finally, we take the limits $N \to \infty$ and possibly $\lambda \to 0$ in order to sharpen the results.

A population state x will now refer to the number of agents currently playing strategy X. The fraction of such agents is denoted by $k = x/N$. Learning is taken to be an infrequent occurrence compared with the play of the game. At the end of each period of length τ, a mental bell rings inside each player's head, an event I refer to as "receiving the learn draw." Let the units in which time is measured be chosen so that the probability of receiving the learn draw in a period of length τ is given by τ.

Learn draws are independent across agents and across time. An agent who does not receive the learn-draw retains her current strategy, while an agent receiving the learn draw potentially changes strategies.

As τ approaches 0, occurrences in which more than one agent receives the learn draw in a single period will be very rare. As a result, the system can be described in terms of the probabilities that, when a learn draw is received, the number of agents currently playing strategy X increases or decreases by one. Let $r_{(\lambda,N)}(x)$ be the probability that, given population size N and mutation rate λ, and given that a single player (and only a single player) receives the learn draw, the result is to cause a player currently playing strategy Y to switch to X. Hence $r_{(\lambda,N)}(N) = 0$. Similarly, let $\ell_{(\lambda,N)}(x)$ be the probability that, given a single player receives the learn draw, the result is to cause a player currently playing X to switch to Y. Hence $\ell_{(\lambda,N)}(0) = 0$.

The parameter $\lambda \geq 0$ that appears in $r_{(\lambda,N)}(x)$ and $\ell_{(\lambda,N)}(x)$ will be interpreted as the rate of mutation. For example, this is the equivalent of the mutation rate in the models of Young [250] and of Kandori, Mailath, and Rob [128], where a mutation causes an agent to abandon her current strategy in order to be randomly assigned to another one. More generally, mutations may capture a variety of minor disturbances, many of which modelers would often suppress in the belief that they are too small to be relevant. Because the focus will sometimes be on what happens as $\lambda \to 0$, it is assumed that $r_{(\lambda,N)}(x)$ and $\ell_{(\lambda,N)}(x)$ are continuous on the right at $\lambda = 0$. I refer to $r_{(0,N)}(x)$ and $\ell_{(0,N)}(x)$ as the "selection" or "learning process."

We assume that the learning process satisfies assumptions 9.1.1–9.1.5, 9.2, and 9.3.1–9.3.3.

Assumption 9.1.1 $r_{(0,N)}(0) = \ell_{(0,N)}(N) = 0$;

Assumption 9.1.2 $\left(x \in \{1, \ldots, N - 1\} \text{ and } \lambda > 0\right) \Rightarrow r_{(\lambda,N)}(x) > h_N(x) > 0$;

Assumption 9.1.3 $\left(x \in \{1, \ldots, N - 1\} \text{ and } \lambda > 0\right) \Rightarrow \ell_{(\lambda,N)}(x) > h_N(x) > 0$;

Assumption 9.1.4 $\left(x \in \{1, \ldots, N - 1\} \text{ and } \lambda > 0\right) \Rightarrow 0 < h \leq \frac{r_{(\lambda,N)}(x)}{\ell_{(\lambda,N)}(x)} \leq H < \infty$;

Assumption 9.1.5 $\lim_{\lambda \to 0} \frac{r_{(\lambda,N)}(0)}{\lambda} \in (h, H)$; $\lim_{\lambda \to 0} \frac{\ell_{(\lambda,N)}(N)}{\lambda} \in (h, H)$,

where $h_N(x)$ is a positive-valued function on $\{1, \ldots, N - 1\}$ and h and H are constants. Assumption 9.1.1, which asserts that the learning process alone cannot cause an agent to switch to a strategy not already present in the population, is common in biological models, where changes in the composition of the population are caused by death and reproduction. It may be appropriate in a learning context if learning is driven primarily by imitation; it will not hold for best-response learning. This assumption could be easily relaxed, but it helps focus attention on the role of mutations in the model. Assumptions 9.1.2–9.1.3 are the essence of the Muddling model. By ensuring that both $r_{(0,N)}(x)$ and $\ell_{(0,N)}(x)$ are positive (except in the pure population states $x = 0$ and $x = N$), they ensure that the learning process may either increase or decrease the number of agents playing strategy X, and hence may either move agents toward or away from best replies, in any state in which both strategies are present in the population. Assumption 9.1.4 ensures that the muddling agents are not too close to being maximizers, in the sense that the probability of moving in the direction of a best response cannot be arbitrarily large compared to the probability of moving away from a best response. It is important here that h and H do not depend on λ or N, so that the model does not collapse to a maximizing model as the probability of a mutation gets small or the population gets large. Assumption 9.1.5, which states that mutations can push the population away from a state in which all agents play the same strategy, ensures that the probability of switching away from a monomorphic state is of the same order of magnitude (for small mutation probabilities) as the probability of a mutation. The interpretation here is that a single mutation suffices to introduce a new strategy into a monomorphic population.

Assumption 9.2 There exist functions $r_\lambda(k)$ and $\ell_\lambda(k)$ that are continuous in λ and k for $0 \leq \lambda \leq 1$ and $0 \leq k \leq 1$ such that

(i) $r_{(\lambda,N)}(kN) = r_\lambda(k) + O(\frac{1}{N})$;

(ii) $\ell_{(\lambda,N)}(kN) = \ell_\lambda(k) + O(\frac{1}{N})$.

Assumption 9.2 is the requirement that as the population gets large, the behavior of an agent who has received the learn draw depends only on the agent's current strategy and the *fraction* of the population playing X. This reflects the fact that as the population gets large, the fraction of an agent's opponents playing a particular strategy and the fraction of the population playing that strategy are nearly identical, causing sampling effects to disappear.

Assumption 9.3.1 $0 < k < 1 \Rightarrow \left(r_0(k) > 0 \text{ and } \ell_0(k) > 0\right)$;

Assumption 9.3.2 $\pi_X(k) > \pi_Y(k) \Leftrightarrow r_0(k) > \ell_0(k)$;

Assumption 9.3.3 $\pi_X(k) < \pi_Y(k) \Leftrightarrow r_0(k) < \ell_0(k)$.

Assumption 9.3.1 ensures that the propensity for agents to switch strategies does not disappear as the population grows large. Assumptions 9.3.2–9.3.3 require that the learning process be always more likely to move the system in the direction of a best reply than away from it. This builds in a link between payoffs and strategy selection, though the muddling nature of the learning ensures that this is an imperfect link.[3]

An example of a muddling model satisfying assumptions 9.1.1–9.1.5, 9.2, and 9.3.1–9.3.3 is the Aspiration and Imitation model of chapter 3. Binmore and Samuelson [24] work with a more general version of the Aspiration and Imitation model, which again satisfies these same assumptions, in which the distribution F of the random variable $\tilde{R}$ that perturbs payoffs is assumed to be log-concave rather than Uniform.

9.2   Dynamics

Stationary Distribution

Examining the ultralong-run behavior of the Muddling model requires a study of the stationary distribution of the system. Fix the game $\hat{G}$. For a fixed set of values of the parameters τ, λ, and N, we then have a homogeneous Markov process $\Gamma_{(\lambda,N,\tau)}$ on a finite state space. In addition, the Markov process is irreducible, because assumptions 9.1.2, 9.1.3, and 9.1.5 ensure that for any state $x \in \{0, 1, \ldots N\}$, there is a positive probability both that the Markov process moves to the state $x + 1$ (if $x < N$), where the number of agents playing X is increased by one, and that the process moves to the state $x - 1$ (if $x > 0$), where the number of agents playing X is decreased by one. The following result is then both standard and familiar from propositions 3.2 and 7.1:

Proposition 9.1   The Markov process $\Gamma_{(\lambda,N,\tau)}$ has a unique stationary distribution. The proportion of time to date t spent in each state converges to the corresponding stationary probability almost surely as $t \to \infty$; and the distribution over states at a given time t converges to the stationary distribution as $t \to \infty$.

---

3. Blume [35] examines a model satisfying assumptions 9.3.1–9.3.3, with agents being more likely (but not certain) to switch to high-payoff strategies and with switching probabilities being smoothly related to payoffs.

Let $\gamma_{(\lambda,N,\tau)}$ be the probability measure given by the stationary distribution, hereafter simply called the "stationary distribution." Then $\gamma_{(\lambda,N,\tau)}(x)$ is the probability attached by the stationary distribution to state x. This chapter studies the distribution $\gamma_{(\lambda,N)}$ obtained from $\gamma_{(\lambda,N,\tau)}$ by taking the limit $\tau \to 0$.

As $\tau \to 0$, the event in which more than one agent receives the learn draw occurs with negligible probability. The model is thus a birth–death process, as studied in Gardiner [97] and by Karlin and Taylor [131, chapter 4]. The following result is standard, where (9.1) is known as the "detailed balance" equation:

Proposition 9.2   Consider states x and $x + 1$. Then the limiting stationary distribution $\lim_{\tau \to 0} \gamma_{(\lambda,N,\tau)} = \gamma_{(\lambda,N)}$ exists and satisfies

$$\frac{\gamma_{(\lambda,N)}(x+1)}{\gamma_{(\lambda,N)}(x)} = \frac{r_{(\lambda,N)}(x)}{\ell_{(\lambda,N)}(x+1)}. \tag{9.1}$$

To interpret (9.1), consider a game with two strict Nash equilibria. Let k^* be the probability attached to X by the mixed-strategy Nash equilibrium of the game and let $x^*/N \equiv k^*$. Notice that x^* need not be an integer, in which case there will be no population state that exactly corresponds to the mixed-strategy equilibrium. Then if $x > x^*$, we must have $\gamma_{(\lambda,N)}(x+1) > \gamma_{(\lambda,N)}(x)$ if λ is sufficiently small and N large because strategy X must be a best reply here, and hence assumption 9.3.2 gives $r_0(x/N) > \ell_0(x/N)$; and assumptions 9.2 and 9.3.1 can then be invoked to conclude $r_{(\lambda,N)}(x) > \ell_{(\lambda,N)}(x+1)$. Hence the stationary distribution $\gamma_{(\lambda,N)}$ increases on $[x^*, N]$. Similarly, from assumption 9.3.3, $\gamma_{(\lambda,N)}(x+1) < \gamma_{(\lambda,N)}(x)$, and $\gamma_{(\lambda,N)}(x)$ must decrease on $[0, x^*]$.[4] The graph of γ therefore reaches maxima at the endpoints of the state space. These endpoints correspond to the strict Nash equilibria of the game at which either all agents play X or all agents play Y. Its minimum is achieved at x^*.[5]

Convergence

How long is the ultralong run, and is it too long? One way of answering this is to compare the convergence properties of the Muddling

---

4. The precise statement here is that for fixed $\epsilon > 0$, there is sufficiently large N and small λ such that $\gamma_{(\lambda,N)}$ increases on $[x^* + \epsilon, N - \epsilon]$ and decreases on $[\epsilon, x^* - \epsilon]$, and whose minimum and maximum on $[x^* - \epsilon, x^* + \epsilon]$ differ by less than ϵ.
5. The convenience of the detailed balance equation (9.1) is available because attention has been restricted to two-player 2×2 games. Extending the analysis to larger games will require new techniques, though some of the basic ideas will reappear.

model with those of Kandori, Mailath, and Rob's model [128]. To do so, the population size N will be fixed and the limit as the probability of a mutation λ gets small will be examined. We shall limit our attention to the commonly studied case of a game with two strict Nash equilibria.

Let $\Psi_{(\lambda,N)}$ be the transition matrix of the Kandori, Mailath, and Rob model, given mutation rate λ and population size N, and let $\psi_{(\lambda,N)}$ be its stationary distribution. Ellison [74] introduces the following measure:

$$\sup_{\psi^0} \limsup_{t\to\infty} \| \psi^0 [\Psi_{(\lambda,N)}]^t - \psi_{(\lambda,N)} \|^{1/t},$$

where ψ^0 is the initial distribution. This is a measure of the distance between the distribution at time t (given by $\psi^0[\Psi_{(\lambda,N)}]^t$) and the stationary distribution (given by $\psi_{(\lambda,N)}$).

Kandori, Mailath, and Rob [128] concentrate their attention on the limiting case of arbitrarily small mutation rates. Ellison [74] shows that there exists a function $h_\psi : \mathbb{R} \to \mathbb{R}$ such that

$$1 - \sup_{\psi^0} \limsup_{t\to\infty} \| \psi^0 [\Psi_{(\lambda,N)}]^t - \psi_{(\lambda,N)} \|^{1/t} = h_\psi(\lambda^z) \sim \lambda^z, \tag{9.2}$$

where z is the minimum number of an agent's opponents that must play the risk-dominant equilibrium strategy in order for this to be a best reply for the agent in question.[6] To see what this implies for the length of time required to approach the stationary distribution, suppose we are interested in the length of time required for the Kandori, Mailath, and Rob model to be within η of its stationary distribution or, equivalently, for $\| \psi^0[\Psi_{(\lambda,N)}]^t - \psi_{(\lambda,N)} \| \le \eta$. Taking $T_\psi(\eta)$ as our measure of the length of time required for the model be within η of its stationary distrubution, (9.2) ensures that for sufficiently large t and small η, we have the approximation $\eta = (1 - h_\psi(\lambda^z))^{T_\psi(\eta)}$ and hence

$$T_\psi(\eta) = \frac{\ln \eta}{\ln(1 - h_\psi(\lambda^z))}. \tag{9.3}$$

As the mutation rate λ gets small, the denominator in (9.3) will thus approach zero quickly, causing the expected waiting time $T_\psi(\eta)$ to grow rapidly.

We now seek an analogous measure of the rate at which the Muddling model converges. To compare the continuous-time Muddling model to the discrete model of Kandori, Mailath, and Rob, first fix

---

6. I say, "The functions $f(\lambda)$ and $g(\lambda)$ are comparable," and write $f \sim g$ if there exist constants c and C such that for all sufficiently small λ, $c|g(\lambda)| \le |f(\lambda)| \le C|g(\lambda)|$.

the unit in which time is to be measured. This unit of measurement will remain constant throughout the analysis, even as the length of the time periods between learn draws in the Muddling model is allowed to shrink. The question then concerns how much time, measured in terms of the fixed unit, must pass before the probability measure describing the expected state of the relevant dynamic process is sufficiently close to its stationary distribution. To make the models comparable, choose the units in which time is measured so that the episodes in which every agent learns in the Kandori, Mailath, and Rob model occur at each of the discrete times $1, 2, \ldots$, and let the probability of receiving the learn draw in the Muddling model, in a period of length τ, be given by τ. Then in the limit as $\tau \to 0$, the expected number of times in an interval of time of length one (which will contain many very short time periods of length τ) that an agent in the Muddling model learns is then one, matching the Kandori, Mailath, and Rob model.

Recall that $\Gamma_{(\lambda,N,\tau)}$ is the transition matrix for the Markov process of the Muddling model, given mutation rate λ and period length τ.[7] Notice also that as τ decreases, the number t/τ of periods up to time t increases. Binmore and Samuelson [24] prove

Proposition 9.3   There exists a function $h_\gamma(\lambda)$ such that

$$\lim_{\tau \to 0} \left(1 - \sup_{\gamma^0} \| \gamma^0 [\Gamma_{(\lambda,N,\tau)}]^{\frac{t}{\tau}} - \gamma_{(\lambda,N)} \|^{\frac{1}{t-1}} \right) \le h_\gamma(\lambda) \sim \lambda, \tag{9.4}$$

where γ^0 is the initial distribution.

Together, (9.2) and (9.4) imply that for very small values of λ, the Muddling model converges faster than does the Kandori, Mailath, and Rob model. In particular, let $T_\gamma(\eta)$ be the length of time required for the Muddling model to be within η of its stationary distribution. Then we have $\eta \ge (1 - h_\gamma(\lambda))^{T_\gamma(\eta)-1}$ for large t and small η. The counterpart of (9.3) is now

$$T_\gamma(\eta) - 1 \le \frac{\ln \eta}{\ln(1 - h_\gamma(\lambda))}. \tag{9.5}$$

Together, (9.3)–(9.5) give, for small values of λ and large t,

$$\frac{T_\psi(\eta)}{T_\gamma(\eta) - 1} \ge \frac{\ln(1 - h_\gamma(\lambda))}{\ln(1 - h_\psi(\lambda^z))} \approx \frac{h_\gamma(\lambda)}{h_\psi(\lambda^z)} \sim \frac{1}{\lambda^{z-1}}. \tag{9.6}$$

---

7. The matrix $\Gamma_{(\lambda,N,\tau)}$ depends on τ because the probability of an agent receiving the learn draw in a given period depends on the period length.

Condition (9.6) implies that if, for example, $N = 100$ and $z = 33$, so that $\frac{1}{3}$ of one's opponents must play the risk-dominant strategy in order for it to be a best reply, then it will take $1/\lambda^{32}$ times as long for the Kandori, Mailath, and Rob model to be within η of its stationary distribution as it takes the Muddling model. Ellison [74] obtains a similar comparison for the Kandori, Mailath, and Rob model and his "two neighbor" matching model. Ellison notes that if $N = 100$ and $z = 33$, then halving the mutation rate causes his two–neighbor matching model (and hence the Muddling model) to take about twice as long to converge, while the Kandori, Mailath, and Rob model will take 2^{33} (> 8 billion) times as long to converge.

What lies behind the result established in proposition 9.3? The Kandori, Mailath, and Rob model relies upon mutations to accomplish its transitions between equilibria. For example, the stationary distribution may put all of its probability on state 0, but the initial condition may lie in the basin of attraction of state N. Best-reply learning then takes the system immediately to state N, and convergence requires waiting until the burst of z simultaneous mutations, required to jump over the basin of attraction of N and reach the basin of attraction of 0, becomes a reasonably likely event. Because the probability of such an event is of the order of λ^z, this requires waiting a very long time when the mutation rate is small. In contrast, the Muddling model requires mutations only to escape boundary states (see assumptions 9.1.1–9.1.5). Once a single mutation has allowed this escape (cf. assumption 9.1.5), then the noisy learning dynamics can allow the system to "swim upstream" out of its basin of attraction.[8] The probability of moving from state N to state 0 is given by $\prod_{x=N}^{1} \ell_{(\lambda,N)}(x)$. When mutation rates are small, the learning dynamics proceed at a much faster rate than mutations occur, so that only one term in this expression ($\ell_{(\lambda,N)}(N)$) is of order λ. Convergence then requires waiting only for a single mutation, rather than z simultaneous mutations, and hence relative waiting times differ by a factor of λ^{z-1}.

The difference in rates of convergence for these two models will be most striking when the mutation rate is very small, and examining an arbitrarily small mutation rate focuses attention on the worst possible case for the Kandori, Mailath, and Rob [128] model. Binmore, Samuelson, and Vaughan [28] present an example in which $N = 100$, $z = 33$,

---

8. Fudenberg and Harris [85] make a similar distinction, including the "swimming upstream" analogy.

and $\lambda = .001$. The expected waiting time in the Kandori, Mailath, and Rob model is approximately 1.7×10^{72}, while that of the Muddling model is approximately 5,000. However, waiting times are likely to be much closer for larger mutation rates, because increasing λ makes the Kandori, Mailath, and Rob model noisier, reducing its waiting time. The waiting times are also likely to be closer if we force the noise in the Muddling model process to become negligible. This comparison thus cannot be viewed as an indictment of the Kandori, Mailath, and Rob model, but rather as an illustration of an important point: incorporating nonnegligible noise into a model can hasten its convergence. Even if unexplained, exogenously determined perturbations (mutations) are to be treated as negligible, expected waiting times can still be short if noise appears in the learning process itself.

This examination of waiting times has held the population size N fixed. The Muddling model yields sharp equilibrium selection results as $N \to \infty$, but need not do so for small values of N. In addition, the expected waiting time diverges in the Muddling model as N increases, becoming arbitrarily long as N gets large.[9] There accordingly remains plenty of room for skepticism as to the applicability of ultralong-run analyses based on examining stationary distributions. In particular, the Muddling model obtains crisp equilibrium selection results only at the cost of long waiting times, and the finding that waiting times can be short must be tempered by the realization that the resulting stationary distribution may give noisy equilibrium selection results. In many cases, however, a population that is not arbitrarily large and a stationary distribution that allocates probability to more than one state may be the most appropriate model, even though it does not give unambiguous equilibrium selection results.[10]

9.3 Equilibrium Selection

We turn now to equilibrium selection, concentrating on the case of large populations and small mutation rates. In particular, we begin

9. Hence convergence in this model is not fast in the second sense that Ellison [74, pp. 1060–1063] discusses because waiting times in the Muddling model do not remain bounded as N gets large.

10. The population size N often need not be very large before most of the mass of the stationary distribution is attached to a single state. In the Binmore, Samuelson, and Vaughan [28] example, where $N = 100$, $z = 33$, and $\lambda = .001$, the stationary distribution places more than .97 probability on states in which at most .05 of the population plays strategy X.

with the limiting stationary distribution of the Markov process as $\tau \to 0$ and then study the limits $N \to \infty$ and $\lambda \to 0$. The order in which these two limits are taken is one of the issues to be examined. Of these two limiting operations, I consider restricting attention to small mutation rates to be especially unrealistic. Allowing the mutation rate to be bounded away from zero complicates the analysis but affects neither the techniques nor the basic nature of the results.[11] I consider the assumption of a large population to be more realistic of the two limits in many applications, but it clearly does not apply to all cases of interest. It is assumed throughout this section that assumptions 9.1.1–9.1.5, 9.2, and 9.3.1–9.3.3 hold.

Two Strict Nash Equilibria

First assume that $A > B$ and $D > C$, so that the game $\hat{G}$ has two strict Nash equilibria. As in the previous section, let $\gamma_{(\lambda,N)}$ denote the limiting stationary distribution of the Markov process on $\{0, 1, \ldots, N\}$ as $\tau \to 0$. Abusing notation, let $\gamma_{(\lambda,N)}$ also denote the corresponding Borel measure on $[0, 1]$. Thus, for an open interval $A \subset [0, 1]$, $\gamma_{(\lambda,N)}(A)$ is the probability of finding the system at a state x with $x/N \in A$. To avoid a tedious special case, assume

$$\int_0^1 (\ln r_0(k) - \ln \ell_0(k))dk \neq 0, \tag{9.7}$$

where assumptions 9.1.4 and 9.2 ensure that the integral exists. Binmore and Samuelson [24] prove

Proposition 9.4 Let (9.7) hold. Then there exists a unique Borel probability measure γ^* on $[0, 1]$ with

$$\lim_{N\to\infty} \lim_{\lambda\to 0} \gamma_{(\lambda,N)} = \lim_{\lambda\to 0} \lim_{N\to\infty} \gamma_{(\lambda,N)} = \gamma^*,$$

where the limits refer to the weak convergence of probability measures. In addition, $\gamma^*(\{0\}) + \gamma^*(\{1\}) = 1$.

This result states that, in the limit as mutation probabilities get small and the population gets large (in any order), the stationary

---

11. Consider, for example, the case of two strict Nash equilibria. If the mutation rate is positive, then taking the limit as the $N \to \infty$ produces a stationary distribution that concentrates all of its probability near one of the strict Nash equilibria, being closer as the mutation rate is smaller. The criterion for which equilibrium is "selected" in this way is a variant of (9.8), with the limits on the integral being adjusted to account for the positive mutation rate.

distribution of the Markov process attaches probability only to the two pure-strategy equilibria. In "generic" cases, for which (9.7) holds, probability will be attached to only one of these equilibria, which I refer to as the "selected equilibrium."

Which Equilibrium?

A number of papers have recently addressed the problem of equilibrium selection in symmetric 2×2 games. Young [250] and Kandori, Mailath, and Rob [128] are typical in finding that the risk-dominant equilibrium is always selected. Robson and Vega Redondo [193] offer a model in which the payoff-dominant equilibrium is always selected. However, condition (9.8) below provides a criterion showing that the Muddling model sometimes selects the payoff-dominant equilibrium and sometimes selects the risk-dominant equilibrium.

Proposition 9.5.1   The selected equilibrium will be (X, X) $[(Y, Y)]$ if

$$\int_0^1 \ln(r_0(k) - \ln \ell_0(k))dk \quad > \quad [<] \ 0 . \tag{9.8}$$

Proposition 9.5.2   The payoff-dominant equilibrium in game $\hat{G}$ can be selected even if it fails to be risk-dominant.

Proposition 9.5.1 is established in the course of proving proposition 9.4. In particular, proposition 9.4 ensures that only states 0 and N, which correspond to the two strict Nash equilibria of the game, receive positive probability in the limit of the stationary distribution as $\lambda \to 0$ and $N \to \infty$. To then see why we can expect the integral (9.8) to determine which of these two equilibria is selected, recall that the stationary distribution satisfies the detailed balance equation given by (9.1). This in turn implies that the probabilities attached to states 0 and N are given by

$$\frac{\gamma_{(\lambda,N)}(N)}{\gamma_{(\lambda,N)}(0)} = \prod_{x=0}^{N-1} \frac{r_{(\lambda,N)}(x)}{\ell_{(\lambda,N)}(x+1)} .$$

Now, taking logarithms of this expression and taking limits gives

$$\lim_{\lambda \to 0} \ln \frac{\gamma_{(\lambda,N)}(N)}{\gamma_{(\lambda,N)}(0)} = \sum_{x=0}^{N-1} \{\ln r_{(0,N)}(x) - \ln \ell_{(0,N)}(x+1)\},$$

and hence

$$\lim_{N\to\infty} \lim_{\lambda\to 0} \frac{1}{N} \ln \frac{\gamma_{(\lambda,N)}(N)}{\gamma_{(\lambda,N)}(0)} = \int_0^1 (\ln r_0(k) - \ln \ell_0(k)) dk, \tag{9.9}$$

which in turn leads to (9.8).

To establish proposition 9.5.2, consider the Aspiration and Imitation model. Let

$$\Delta = 0, \quad A = 2, \quad B = 1, \quad D = 0, \quad C = -1.$$

Then neither of the two pure-strategy Nash equilibria, given by (X, X) and (Y, Y), risk-dominates the other, but (X, X) is the payoff-dominant equilibrium. In order to compute the outcome, let F be a Uniform distribution on the interval $[-2, 2]$. Then the probability of abandoning a strategy is linear in the expected payoff of the strategy, with

$$g(2) = 0, \quad g(1) = \tfrac{1}{4}, \quad g(0) = \tfrac{1}{2}, \quad g(-1) = \tfrac{3}{4}.$$

Inserting these probabilities in the expressions for transition probabilities for the Aspiration and Imitation model given by (3.8)–(3.11), taking the limits $\lambda \to 0$ and $N \to \infty$, and inserting in (9.8) gives

$$\frac{\gamma^*(1)}{\gamma^*(0)} = \lim_{N\to\infty} N \int_0^1 \left(\ln g(\pi_Y(k)) - \ln g(\pi_X(k)) \right) dk$$

$$= \lim_{N\to\infty} N \int_0^1 \left(\ln(\tfrac{1}{4}k + \tfrac{1}{2}(1-k)) - \ln(0k + \tfrac{3}{4}(1-k)) \right) dk$$

$$= \lim_{N\to\infty} \left(\frac{4}{3} \right)^N, \tag{9.10}$$

ensuring that (X, X) is selected. The game can then be perturbed slightly to make (Y, Y) risk-dominant while still keeping (X, X) payoff-dominant and without altering the fact that (X, X) is selected.

Why does this result differ from that of Kandori, Mailath, and Rob [128], whose model selects the equilibrium with the larger basin of attraction under best-reply dynamics, namely, the risk-dominant equilibrium? In the perturbed version of the game considered in the previous proof, the equilibrium (X, X) has a basin of attraction smaller than (Y, Y)'s, but in (X, X)'s basin the probability of switching away from strategy X, relative to the corresponding probability for strategy Y, is exceptionally small, being nearly zero for states in which nearly all agents play X. This makes it very difficult to leave (X, X)'s basin, and yields a selection in which all agents play X. Only the size of the basin of attraction matters in Kandori, Mailath, and Rob, while in the Muddling model the strength of the learning flows matters as well.

Best-Response Dynamics

If it is the relative strengths of the learning flows that cause the Muddling model to sometimes select the payoff-dominant equilibrium, then the model should be more likely to select the risk-dominant equilibrium if the learning process is closer to best-response learning, so that "strength of flow" issues are less important. This can be confirmed.

Let $A > B$ and $D > C$, and let $k^*/N > \frac{1}{2}$, where k^* is the probability attached to X by the mixed-strategy equilibrium, so that there are two strict Nash equilibria, with (Y, Y) being the risk-dominant equilibrium. Fix $r_0(k)$ and $\ell_0(k)$ satisfying assumptions 9.1.1–9.1.5, 9.2, and 9.3.1–9.3.3. Then let

$$\tilde{r}_0(k) = \phi B_X(k) + (1 - \phi)r_0(k)$$

$$\tilde{\ell}_0(k) = \phi B_Y(k) + (1 - \phi)\ell_0(k),$$

where $B_X(k)$ equals 1 if X is a best response ($k > k^*$), and zero otherwise, and $B_Y(k)$ equals one if Y is a best response ($k < k^*$), and zero otherwise. Then $\tilde{r}_0(k)$ and $\tilde{\ell}_0(k)$ are a convex combination of the best-response dynamics and $r_0(k)$ and $\ell_0(k)$. As ϕ increases to unity, $\tilde{r}_0$ and $\tilde{\ell}_0$ approach best-response dynamics. Binmore and Samuelson [24] prove:

Proposition 9.6 If $r_0(k)$ and $\ell_0(k)$ satisfy assumptions 9.1.1–9.1.5, 9.2, and 9.3.1–9.3.3, then the selected equilibrium in game $\hat{G}$ is the risk-dominant equilibrium for any convex combination of the best-response dynamics and $r_0(k)$ and $\ell_0(k)$ that puts sufficient weight on the former.

Random Imitation Dynamics

Best-response dynamics have the property that strategy adjustments are precisely determined by payoffs. At another end of the spectrum are dynamics in which strategy adjustments have nothing to do with payoffs. Suppose that those who receive the learn draw imitate a randomly chosen opponent, regardless of payoff considerations. In the limiting case as $N \to \infty$ and $\lambda \to 0$, the probability that such imitation increases or decreases the number of agents playing X by one, given that the proportion playing X is currently k, is then $k(1-k)$.[12]

Now suppose that this random imitation process governs strategy adjustments with probability θ, while with probability $1 - \theta$ adjustments are given by $r_{(\lambda,N)}(x)$ and $\ell_{(\lambda,N)}(x)$, which satisfy assumptions 9.1.1–9.1.5, 9.2, and 9.3.1–9.3.3. Then in the limiting case of a large population and small mutation rate, transition probabilities are given by

$$\tilde{r}_0(k) = \theta k(1-k) + (1-\theta)r_0(k)$$

$$\tilde{\ell}_0(k) = \theta k(1-k) + (1-\theta)\ell_0(k).$$

If $\theta = 1$, then the process is driven *entirely* by random imitation, and the evolution of the system has nothing to do with the payoffs in the game $\hat{G}$. If $\theta = 0$, then there is no random imitation. I refer to θ as the "level of random imitation."

The process of random imitation is reminiscent of the concept of background fitness in biology, where it is the rate at which a strategy reproduces and grows due to "background" considerations lying outside the game in question. Chapter 2 presented versions of the replicator dynamics in which a background fitness term appears. There are several alternative interpretations of random imitation in an economic setting. For example, agents may be occasionally withdrawn from the game-playing population to be replaced by inexperienced novices who choose incumbents to imitate. Alternatively, there may be some probability that an agent's thought processes are currently congested with other considerations and hence decisions to abandon strategies and imitate others are made without any reference to payoffs. It may be that the agent is actually participating in a large number of similar games using the same strategy for each, as in the model of Carlsson and van Damme [61]. Strategy revisions will then often be determined by what happens in games other than the game actually under study and may again take the form of agents in this game randomly abandoning their strategy to imitate that of another.

---

12. The probability of increasing the number of agents playing X by one is the probability that the agent who dies is playing strategy Y (given by $(N-x)/N = 1-k$ in the limit as N gets large) and the probability that the agent giving birth plays strategy X (given by $x/N = k$ for the limiting case of large N and small λ), giving $k(1-k)$.

Proposition 9.7 If the level of random imitation θ is sufficiently large, then the equilibrium (X, X) is selected if

$$\int_0^1 \frac{1}{k(1-k)}(\ell_0(k) - r_0(k))dk < 0, \tag{9.11}$$

and (Y, Y) is selected if (9.11) is positive.

Proof of Proposition 9.7 Let (X, X) and (Y, Y) be Nash equilibria, with (Y, Y) being risk dominant. From (9.8), the selected equilibrium will be (X, X) if

$$\int_0^1 \Big\{ \ln(k(1-k)\theta + (1-\theta)r_0(k))$$

$$- \ln(k(1-k)\theta + (1-\theta)\ell_0(k)) \Big\} dk > 0. \tag{9.12}$$

As $\theta \to 1$, the left side of (9.12) approaches zero. Then consider the first derivative of the left side of (9.12). A negative derivative ensures that (9.12) approaches zero from above, so that (9.12) is positive and hence the payoff-dominant equilibrium is selected for θ near 1. Evaluated at $\theta = 1$, this derivative is given by the left side of (9.11). The selected equilibrium, for large background fitness, is thus (X, X) if (9.11) is negative and (Y, Y) if (9.11) is positive.   □

What is the interpretation of condition (9.11)? It is straightforward to evaluate (9.11) for the case of the Aspiration and Imitation model, finding

Corollary 9.1 In the Aspiration and Imitation model, (9.11) is negative and hence the risk-dominant equilibrium of game $\hat{G}$ is selected, for sufficiently large θ.

This provides another condition for the risk-dominant equilibrium to be selected. In the case of the Aspiration and Imitation model, the condition is that changes in strategy must be driven primarily by random imitation and not payoffs in the game. Hence the risk-dominant equilibrium is selected in relatively unimportant games—those whose payoffs have little to do with agents' behavior.

No Pure-Strategy Equilibria

The previous findings establish equilibrium selection results for the case of two strict Nash equilibria. These can be contrasted with the case of games in which $B > A$ and $C > D$, so that there is a unique, mixed-strategy Nash equilibrium. Then, as indicated in chapter 3 for the case of the Aspiration and Imitation model, an argument analogous to that of proposition 9.4 yields

Proposition 9.8   Let k^* be the probability attached to X in the mixed-strategy equilibrium. Then $\lim_{\lambda \to 0} \lim_{N \to \infty} \gamma_{(\lambda,N)}([k_1, k_2]) = 0$ if $k^* \notin [k_1, k_2]$. However, $\lim_{\lambda \to 0} \gamma_{(\lambda,N)}(\{0\}) + \lim_{\lambda \to 0} \gamma_{(\lambda,N)}(\{1\}) = 1$.

The order of limits makes a difference in this case. If mutation rates are first allowed to approach zero, then the ultralong-run dynamics are driven by the possibility of accidental extinction coupled with the impossibility of recovery, attaching probability only to the two nonequilibrium states in which either all agents play X or all agents play Y. If the population size is first allowed to get large, then accidental extinctions are not a factor and the long-run outcome is the mixed-strategy equilibrium. My inclination here is to regard the latter as the more useful model.

9.4   Risk Dominance

The results of the previous section involve changes in the learning rule. In this section the learning rule is fixed, and we examine changes in the payoffs of the game. This provides ultralong-run comparative statics to complement the long-run comparative statics examined in chapter 6.

It is immediately apparent that some additional assumptions are required to establish comparative static results. Assumptions 9.1.1–9.1.5, 9.2, and 9.3.1–9.3.3 are silent on the question of how changes in payoffs affect the learning dynamics as long as the inequalities in assumptions 9.3.1–9.3.3 are satisfied. Accordingly, let us consider the Aspiration and Imitation model, investigating games with two strict Nash equilibria ($A > B$ and $D > C$), with the question being when the risk-dominant equilibrium will be selected. First, we consider the case in which there is no conflict between payoff and risk dominance:

Proposition 9.9   If the payoff-dominant equilibrium in game $\hat{G}$ is also risk-dominant, then the payoff-dominant equilibrium is selected.

Proof of Proposition 9.9   In the Aspiration and Imitation model, the criterion given by (9.8) for the selection of equilibrium (X, X) becomes

$$\int_0^1 \left(\ln F(\Delta - \pi_Y(k)) - \ln F(\Delta - \pi_X(k)) \right) dk > 0. \tag{9.13}$$

Let (X, X) and (Y, Y) be risk-equivalent in game $\hat{G}$, so that $A + C = B + D$, and let $A = D$. Then from the linearity of $\pi_X(k)$ and $\pi_Y(k)$, we have that (9.13) holds with equality. Now make (X, X) the payoff-dominant equilibrium by increasing A and decreasing C so as to preserve $A + C = B + D$ (and hence to preserve the risk-equivalence of the two strategies). Because F is Uniform, this adjustment causes $\int_0^1 \ln F(\Delta - \pi_X)dk$ to decrease. The payoff-dominant equilibrium (X, X) is then selected. Next, note that adding a constant to A and C or subtracting a constant from D and B, so as to also make (X, X) risk-dominant, increases the function $\ln F(\Delta - \pi_Y(k)) - \ln F(\Delta - \pi_X(k))$ on $[0, 1]$, which strengthens the inequality in (9.13) and hence preserves the result that the payoff-dominant equilibrium is selected.   □

Now consider cases where the payoff and risk dominance criteria conflict. We have already seen that sometimes the payoff-dominant equilibrium will be selected and sometimes the risk-dominant equilibrium will be selected. The question of interest is how the likelihood of choosing the risk-dominant equilibrium in the game $\hat{G}$ varies with the payoffs in the game. To pose this question precisely, let k^* be the probability attached to X in the mixed strategy equilibrium. Let (Y, Y) be the risk-dominant equilibrium (so $k^* > \frac{1}{2}$) but let (X, X) be payoff-dominant. Let π^* be the payoff in game $\hat{G}$ from the mixed-strategy equilibrium. We can then consider variations in the payoffs A, B, C, D that leave k^* and π^* unchanged. For example, we can consider a decrease in D accompanied by an increase in B calculated so as to preserve k^* and π^*, as illustrated in figure 9.2. We thus restrict our attention to variations in the payoffs A, B, C, and D for which $C = C(A)$ and $B = B(D)$, where $(1 - k^*)C(A) + k^*A = (1 - k^*)B(D) + k^*D = \pi^*$.

Proposition 9.10   If the payoff-dominant equilibrium is selected given payoffs A, B, C, and D, with mixed-strategy equilibrium k^* and mixed-strategy equilibrium payoff π^*, and if $D > B$, then the payoff-dominant equilibrium is also selected for any payoffs A, B', C, and D' that preserve k^* and π^* and for which $D' \in [\pi^*, D]$.

Proposition 9.10 tells us that if $D > B$ and the payoff-dominant equilibrium is selected, then the payoff-dominant equilibrium continues to be selected as D is decreased (and B increased), at least until $D = \pi^*$. The proof establishes that the payoff-dominant equilibrium will continue to be selected for at least some range of values of $D < B$. Hence, if the payoff-dominant equilibrium is selected for any value $D > \pi^*$, then it is selected for an interval of values of D that includes π^* in its inte-

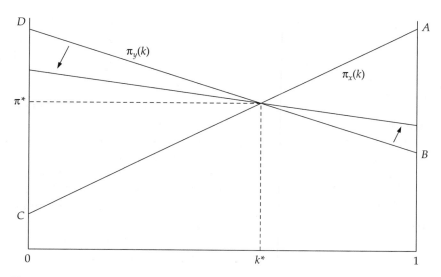

Figure 9.2
Payoff variations

rior. If the payoff-dominant equilibrium is not selected for any value of $D > \pi^*$, then it may still be selected for some value of $D < \pi^*$. Movements in the payoffs to strategy Y that increase B and decrease D then make the payoff-dominant equilibrium more likely to be selected. The best case for the payoff-dominant equilibrium occurs when $D = B$ or $D < B$. The payoff-dominant equilibrium is thus favored by reducing the variation in payoffs to strategy Y or even "inverting" them, so that while (Y, Y) is an equilibrium, the highest reward to strategy Y is obtained if the opponent plays X.

Proof of Proposition 9.10 Fix $k^* > 1/2$. Fix A and hence $C(A)$. If we set $D = A$, then proposition 9.9 ensures that (Y, Y) will be selected, since it is risk dominant and payoff undominated. Now let D decline. Because F is Uniform, (9.13) increases until D reaches π^*, and is increasing at $D = \pi^*$. Hence, if the payoff-dominant equilibrium is selected for any value of $D > \pi^*$, then it is also selected for any smaller value of D that preserves $D \geq \pi^*$. $\qquad\square$

9.5   Discussion

Given proposition 9.10, it is interesting to note that Straub [229] has conducted experiments to investigate the conditions under which risk-dominant and payoff-dominant equilibria are selected in 2×2 symmetric games with two strict Nash equilibria. He finds that the risk-dominant equilibrium is the most common equilibrium played in seven out of eight of the experiments. The exception, in which the payoff-dominant equilibrium appeared, is the only game in which $D < B$. This is consistent with the statement that the payoff-dominant equilibrium is most likely to be selected when $D \leq B$.

Friedman [83] also reports experiments with 2×2 symmetric games having two strict Nash equilibria. Friedman finds that altering a game in which $D < B$ to achieve $D = B$, while preserving the basins of attraction of the Nash equilibria, causes the risk-dominant equilibrium to be selected more often and the payoff-dominant equilibrium to be selected less often. This is consistent with the statement that the risk-dominant equilibrium may be selected for all values $D \geq B$, but there may exist values $D < B$ for which the payoff-dominant equilibrium is selected.

The experiments of Straub [229] and Friedman [83] were designed for other purposes, and are at best suggestive concerning the circumstances under which risk-dominant or payoff-dominant equilibria will appear. More specifically designed experiments are required to provide more precise insights. Suppose such experiments produce the result that setting $D > B$ yields the risk-dominant equilibrium, while reducing D and increasing B, in such a way as to preserve π^* and the basins of attraction, leads to the payoff-dominant equilibrium. Do we have any reason to suspect that the considerations captured by the Muddling model of this chapter have anything to do with the outcome? To the contrary, we probably have some reason to suspect that this chapter's model does not lie behind the outcome. Proposition 9.10 describes the implications of variations in payoffs for the stationary distribution of the Muddling model. The stationary distribution is an ultralong-run phenomenon. We can safely expect the ultralong run to be longer than the number of trials typically allowed in laboratory experiments. We must then question the relevance of the theory to experiments (though we then also have a convenient apology if experimental outcomes do not match the predictions of the theory).

There remains some hope for the theory. The experiences of experimental subjects outside the laboratory will have directed them to strategies that produce sometimes the risk-dominant equilibrium and sometimes the payoff-dominant equilibrium. If the forces driving this process bear a sufficient relationship to those driving the Muddling model, and if the choices subjects must make in the laboratory sufficiently resemble those they make outside as to trigger behavior appropriately suited to the experimental game, then we can expect the theoretical and laboratory results to be related and hence can expect the theory to be relevant. This link between theoretical and experimental outcomes clearly places a great premium on successful experimental design.

On the one hand, it would seem an extraordinary coincidence that a model as simple as the Muddling model could actually hope to describe how people behave. Notice, however, that traditional game theory tells us that when examining a symmetric 2×2 game with two strict Nash equilibria, only the best-reply structure of the game matters for equilibrium selection, that is, only the sizes of the basins of attraction of the two Nash equilibria matter. The evolutionary models of Kandori, Mailath, and Rob [128] and of Young [250] also produce a result where only the best-reply structure matters. The common intuition is that other considerations, such as the magnitudes of payoffs, also matter. The Muddling model directs our attention to one way in which such considerations might matter.

10 Conclusion

What remains to be done? Two areas for further work strike me as particularly important. First, work on the structure of evolutionary models is needed. An evolutionary analysis begins with a model of how players make their choices in games and how they learn that some choices work better than others. This process makes a more explicit appearance in some models than in others, but is always lurking in the background. Learning is a complicated phenomenon even in a stationary environment. An added layer of complexity appears in games because players can be attempting to learn a moving target. We have only very simple and preliminary ideas about what drives learning, and there is much to be done in this respect.

Additional work is also required on developing techniques to analyze learning models. The current practice relies heavily on being able to restrict attention to either very simple games or simple dynamic processes. Much can be learned by examining the replicator dynamics or Markov processes with one-dimensional state variables, but we must also move beyond such models. In connection with this, it is important to note that existing models typically call for players to be repeatedly matched in pairs to play a fixed game. It is very difficult to think of cases in which one is repeatedly matched with different opponents to play a large number of literally identical games. One possibility in this regard is to argue that people are constantly faced with making decisions in a variety of strategic situations that they sort into a relatively small number of basic categories, such as coordination games, bargaining games, prisoners' dilemmas, and so on. The evolutionary process may then operate within these classes of games, with agents applying a single decision rule to each class that is adjusted in light of their experience in such games. While the general outlines of this story seem plausible, it has not been subjected to the scrutiny of a model.

What of the future of evolutionary game theory? I think an analogy with general equilibrium theory is useful. After establishing the existence and basic properties of Walrasian equilibrium, economists turned considerable attention to the interpretation of this equilibrium. How was an equilibrium achieved, who set prices, how stable was an equilibrium, what happened when there was no Walrasian auctioneer? The resulting analysis produced adjustment models, disequilibrium models, stability analyses, strategic models, models with imperfect competition, and a variety of other investigations. But little change has occurred in the way most economists think about general equilibrium theory, where Walrasian equilibrium remains the standard tool.

Game theorists have established the existence and basic properties of Nash equilibrium and its various refinements. Some variant of sequential equilibrium, including perfect Bayesian equilibrium, is the standard concept. The use of game theory has spread throughout applied fields in economics such as industrial organization and international trade. Evolutionary game theory is now probing many of the same questions that general equilibrium theorists considered. One possibility is that evolutionary game theory follows the alternatives to Walrasian general equilibrium theory in having little effect on how economists think about equilibrium. Evolutionary game theory is especially in danger of suffering this fate if it does no more than produce an ever-growing list of alternative models with differing results.

The future for evolutionary game theory is more hopeful if the theory can demonstrate its ability to provide insights that are useful to economists working with game-theoretic models. One first step to such a demonstration would be to show that evolutionary models can be useful in interpreting and predicting the results of laboratory experiments. A complementary step would be work designed to replace the randomly matched pairs and abstract games of most evolutionary models with patterns of interaction that more closely resemble activities in the markets to which the models are applied. The models could then be used as a standard tool in applied work. Notice that this may require different models for different applications, bringing institutional features into the economic analysis that are currently often thought to be important but are commonly ignored. I expect both avenues of inquiry to play important roles in shaping the future of evolutionary game theory.

References

1. L. Anderlini. Communication, computability, and common interest games. Economic Theory discussion paper 159, St. John's College, Cambridge, 1990.

2. James Andreoni and John H. Miller. Auctions with artificial adaptive agents. *Games and Economic Behavior* 10:39–64, 1995.

3. Robert Axelrod. *The evolution of cooperation*. Basic Books, New York, 1984.

4. Dieter Balkenborg. Strictness, evolutionary stability and repeated games with common interests. CARESS working paper 93–20, University of Pennsylvania, 1993.

5. Dieter Balkenborg. An experiment on forward versus backward induction. SFB discussion paper B-268, University of Bonn, 1994.

6. Dieter Balkenborg. Strictness and evolutionary stability. Center for Rationality and Interactive Decision Theory discusssion paper 52, Hebrew University of Jerusalem, 1994.

7. Dieter Balkenborg and Karl Schlag. Evolutionarily stability in asymmetric population games. SFB discussion paper B-314, University of Bonn, 1995.

8. Dieter Balkenborg and Karl Schlag. On the interpretation of evolutionarily stable sets. SFB discussion paper B-313, University of Bonn, 1995.

9. Abhijit Banerjee and Drew Fudenberg. Word-of-mouth learning. Mimeo, Massachusetts Institute of Technology and Harvard University, 1995.

10. Pierpaolo Battigalli. Strategic rationality orderings and the best rationalization principle. Mimeo, Princeton University, 1993.

11. Pierpaolo Battigalli. On rationalizability in extensive games. *Journal of Economic Theory*, 1996. Forthcoming.

12. Jonathan Bendor, Dilip Mookherjee, and Debraj Ray. Aspirations, adaptive learning, and cooperation in repeated games. Mimeo, Indian Statistical Institute, 1994.

13. James Bergin and Bart Lipman. Evolution with state-dependent mutations. *Econometrica* 64:943–956, 1996.

14. Douglas Bernheim. Rationalizable strategic behavior. *Econometrica* 52:1007–1028, 1984.

15. V. Bhaskar. Noisy communication and the evolution of cooperation. Mimeo, Delhi School of Economics, 1995.

16. N. P. Bhatia and G. P. Szegö. *Stability theory of dynamical systems*. Springer, New York, 1970.

17. Patrick Billingsley. *Probability and measure*. Wiley, New York, 1986.

18. Ken Binmore. Modelling rational players: I and II. *Economics and Philosophy* 3:9–55, 4:179–214, 1987–1988.

19. Ken Binmore. *Fun and games*. D. C. Heath, Lexington, MA, 1992.

20. Ken Binmore. *Game theory and the social contract*. Vol. 1. *Playing fair*. MIT Press, Cambridge, MA, 1993.

21. Ken Binmore, John Gale, and Larry Samuelson. Learning to be imperfect: The ultimatum game. *Games and Economic Behavior* 8:56–90, 1995.

22. Ken Binmore, Chris Proulx, Larry Samuelson, and Joe Swierzbinski. Hard bargains and lost opportunities. SSRI working paper 9517, University of Wisconsin, 1995.

23. Ken Binmore and Larry Samuelson. Evolutionary stability in repeated games played by finite automata. *Journal of Economic Theory* 57:278–305, 1992.

24. Ken Binmore and Larry Samuelson. Muddling through: Noisy equilibrium selection. SSRI working paper 9410R, University of Wisconsin, 1993.

25. Ken Binmore and Larry Samuelson. Drift. *European Economic Review* 38:859–867, 1994.

26. Ken Binmore and Larry Samuelson. An economist's perspective on the evolution of norms. *Journal of Institutional and Theoretical Economics* 150:45–63, 1994.

27. Ken Binmore and Larry Samuelson. Evolutionary drift and equilibrium selection. SSRI working paper 9529, University of Wisconsin, 1994.

28. Ken Binmore, Larry Samuelson, and Richard Vaughan. Musical chairs: Modelling noisy evolution. *Games and Economic Behavior* 11:1–35, 1995.

29. Ken Binmore, Avner Shaked, and John Sutton. Testing noncooperative bargaining theory: A preliminary study. *American Economic Review* 75:1178–1180, 1985.

30. K. Binmore, J. Swierzbinski, S. Hsu, and C. Proulx. Focal points and bargaining. *International Journal of Game Theory* 22:381–409, 1993.

31. Jonas Björnerstedt. Experimentation, imitation and evolutionary dynamics. Mimeo, Stockholm University, 1995.

32. Jonas Björnerstedt, Martin Dufwenberg, Peter Norman, and Jörgen Weibull. Evolutionary selection dynamics and irrational survivors. Research Papers in Economics 1994:7 WE, University of Stockholm, 1994.

33. Andreas Blume. Learning, experimentation and long-run behavior in games. Mimeo, University of Iowa, 1994.

34. Andreas Blume, Yong-Gwan Kim, and Joel Sobel. Evolutionary stability in games of communication. *Games and Economic Behavior* 5:547–575, 1993.

35. Lawrence E. Blume. The statistical mechanics of strategic interaction. *Games and Economic Behavior* 5:387–424, 1993.

36. Lawrence E. Blume. How noise matters. Mimeo, Cornell University, 1994.

37. Gary E. Bolton. A comparative model of bargaining: Theory and evidence. *American Economic Review* 81:1096–1136, 1991.

38. Gary E. Bolton and Rami Zwick. Anonymity versus punishment in ultimatum bargaining. *Games and Economic Behavior* 10:95–121, 1995.

39. Immanuel Bomze. Non-cooperative 2-person games in biology: A classification. *International Journal of Game Theory* 15:31–59, 1986.

40. Immanuel Bomze and Reinhard Bürger. Stability by mutation in evolutionary games. *Games and Economic Behavior* 11:145–173, 1995.

41. Immanuel Bomze and Jurgen Eichberger. Evolutive versus naive Bayesian learning. Institute of Statistics and Computer Science technical report 118, University of Vienna, 1992.

42. Immanuel M. Bomze and Benedikt M. Pötscher. *Game theoretical foundations of evolutionary stability*. Springer, Berlin, 1989.

43. Immanuel M. Bomze and Eric van Damme. A dynamical characterization of evolutionarily stable states. CentER for Economic Research discussion paper 9045, Tilburg University, 1990.

44. Immanuel M. Bomze and Jörgen Weibull. Does neutral stability imply Lyapunov stability? *Games and Economic Behavior* 11:173–192, 1995.

45. Tilman Börgers. On the definition of rationalizability in extensive games. Discussion paper 91-22, University College London, 1991.

46. Tilman Börgers. A formalization of "forward induction." Mimeo, University College London, 1992.

47. T. Börgers and R. Sarin. Learning through reinforcement and replicator dynamics. Mimeo, University College London, 1995.

48. T. Börgers and R. Sarin. Naive reincorcement learning with endogenous aspirations. Mimeo, University College London, 1996.

49. Richard T. Boylan. Laws of large numbers for dynamical systems with randomly matched individuals. *Journal of Economic Theory* 57:473–504, 1992.

50. Richard T. Boylan. Evolutionary equilibria resistant to mutations. *Games and Economic Behavior* 7:10–34, 1994.

51. Richard T. Boylan. Continuous approximation of dynamical systems with randomly matched individuals. *Journal of Economic Theory* 66:615–625, 1995.

52. Eva Bruch. The evolution of cooperation in neighborhood structures. Master's thesis, University of Bonn, 1993.

53. Robert R. Bush and Frederick Mosteller. *Stochastic models for learning*. Wiley, New York, 1955.

54. Antonio Cabrales. Stochastic replicator dynamics. Mimeo, Universitat Pompeu Fabra, Barcelona, 1995.

55. Antonio Cabrales and Joel Sobel. On the limit points of discrete selection dynamics. *Journal of Economic Theory* 57:407–419, 1992.

56. David Canning. Convergence to equilibrium in a sequence of games with learning. STICERD discussion paper 89/190, London School of Economics, 1989.

57. David Canning. Average behavior in learning models. *Journal of Economic Theory* 57:442–472, 1992.

58. David Canning. Learning language conventions in common interest signaling games. Department of Economics discussion paper 607, Columbia University, 1992.

59. David Canning. Learning the subgame-perfect equilibrium. Department of Economics discussion paper 608, Columbia University, 1992.

60. Hans Carlsson and Eric van Damme. Equilibrium selection in Stag-Hunt games. In Ken Binmore, Alan Kirman, and Piero Tani, eds., *Frontiers of Game Theory*, pp. 237–254. MIT Press, Cambridge, MA, 1993.

61. Hans Carlsson and Eric van Damme. Global games and equilibrium selection. *Econometrica* 61:989–1018, 1993.

62. Yin-Wong Cheung and Daniel Friedman. Learning in evolutionary games: Some laboratory results. UCSC Economics Department working paper 303, University of California, Santa Cruz, 1994.

63. In-Koo Cho. Stationarity, rationalizability and bargaining. *Review of Economic Studies* 61:357–374, 1994.

64. In-Koo Cho and David M. Kreps. Signaling games and stable equilibria. *Quarterly Journal of Economics* 102:179–221, 1987.

65. Vincent P. Crawford. An "evolutionary" interpretation of Van Huyck, Battalio, and Beil's experimental results on coordination. *Games and Economic Behavior* 3:25–59, 1991.

66. Vincent P. Crawford. Adaptive dynamics in coordination games. *Econometrica* 63: 103–143, 1995.

67. Vincent P. Crawford. Theory and experimentation in the analysis of strategic interaction. Discussion paper 95–37, University of California, San Diego, 1995.

68. Ross Cressman. Evolutionary game theory with two groups of individuals. *Games and Economic Behavior* 11:237–253, 1995.

69. Ross Cressman. Local stability of smooth selection dynamics for normal-form games. Mimeo, Wilfrid Laurier University, 1996.

70. Ross Cressman and Karl H. Schlag. The dynamic (in)stability of backwards induction. Discussion paper B–347, University of Bonn, 1995.

71. R. Dawkins. *The selfish gene*. Oxford University Press, Oxford, 1976.

72. Eddie Dekel and Suzanne Scotchmer. On the evolution of optimizing behavior. *Journal of Economic Theory* 57:392–406, 1992.

73. Jurgen Eichberger, Hans Haller, and Frank Milne. Naive Bayesian learning in 2×2 matrix games. *Journal of Economic Behavior and Organization* 21:69–90, 1993.

74. Glenn Ellison. Learning, local interaction, and coordination. *Econometrica* 61:1047–1072, 1993.

75. Glenn Ellison. A little rationality and learning from personal experience. Mimeo, Massachusetts Institute of Technology, 1993.

76. Glenn Ellison. Basins of attraction and long run equilibria. Mimeo, Massachusetts Institute of Technology, 1995.

77. Dean Foster and Rakesh V. Vohra. Asymptotic calibration. Mimeo, University of Pennsylvania and Ohio State University, 1995.

78. Dean Foster and Rakesh V. Vohra. Calibrated learning and correlated equilibrium. Mimeo, University of Pennsylvania and Ohio State University, 1995.

79. Dean Foster and Peyton Young. Stochastic evolutionary game dynamics. *Journal of Theoretical Biology* 38:219–232, 1990.

80. R. H. Frank. *Passions within reason*. Norton, New York, 1988.

81. M. I. Freidlin and A. D. Wentzell. *Random perturbations of dynamical systems*. Springer, New York, 1984.

82. D. Friedman. Evolutionary games in economics. *Econometrica* 59:637–666, 1991.

83. Daniel Friedman. Equilibrium in evolutionary games: Some experimental results. *Economic Journal* 106:1–25, 1996.

84. James W. Friedman. On entry preventing behavior. In S. J. Brams, A. Schotter, and G. Schwodiauer, eds., *Applied game theory*, pp. 236–253. Physica, Vienna, 1979.

85. Drew Fudenberg and Chris Harris. Evolutionary dynamics with aggregate shocks. *Journal of Economic Theory* 57:420–441, 1992.

86. Drew Fudenberg and David M. Kreps. A theory of learning, experimentation, and equilibrium in games. Mimeo, Stanford University and Massachusetts Institute of Technology, 1988.

87. Drew Fudenberg and David M. Kreps. Learning mixed equilibria. *Games and Economic Behavior* 5:320–367, 1993.

88. Drew Fudenberg and David M. Kreps. Learning in extensive-form games. 1. Self-confirming equilibria. Harvard Institute for Economic Development discussion paper 20, Harvard University, 1994.

89. Drew Fudenberg and David M. Kreps. Learning in extensive-form games. 2. Experimentation and Nash equilibrium. Harvard Institute for Economic Development discussion paper 20, Harvard University, 1994.

90. Drew Fudenberg and David K. Levine. How irrational are subjects in extensive-form games? Economic Theory discussion paper 14, Harvard University, 1993.

91. Drew Fudenberg and David K. Levine. Self-confirming equilibrium. *Econometrica* 61:523–546, 1993.

92. Drew Fudenberg and David K. Levine. Steady state learning and Nash equilibrium. *Econometrica* 61:547–574, 1993.

93. Drew Fudenberg and David K. Levine. Consistency and cautious fictitious play. Harvard Institute of Economic Research discussion paper 21, Harvard University, 1994.

94. Drew Fudenberg and David K. Levine. Conditonal universal consistency. Mimeo, Harvard University and University of California at Los Angeles, 1995.

95. Drew Fudenberg and David K. Levine. Theory of learning in games. Mimeo, Harvard University and University of California at Los Angeles, 1995.

96. Drew Fudenberg and Eric Maskin. The folk theorem in repeated games with discounting and incomplete information. *Econometrica* 54:533–554, 1986.

97. C. W. Gardiner. *Handbook of stochastic methods.* Springer, Berlin, 1985.

98. D. Gauthier. *Morals by agreement.* Clarendon Press, Oxford, 1986.

99. Itzhak Gilboa and Akihiko Matsui. Social stability and equilibrium. *Econometrica* 59:859–868, 1991.

100. Itzhak Gilboa and Akihiko Matsui. A model of random matching. *Journal of Mathematical Economics* 21:185–198, 1992.

101. Itzhak Gilboa and Dov Samet. Absorbent stable sets. Center for Mathematical Studies in Economics and Management Science discussion paper 935, Northwestern University, 1991.

102. Itzhak Gilboa and David Schmeidler. Case-based consumer theory. Mimeo, Northwestern University, 1993.

103. Itzhak Gilboa and David Schmeidler. Case-based decision theory. *Quarterly Journal of Economics* 110:605–640, 1995.

104. Itzhak Gilboa and David Schmeidler. Case-based optimization. *Games and Economic Behavior,* 1996. Forthcoming.

105. Natalie S. Glance and Bernardo A. Huberman. The outbreak of cooperation. *Journal of Mathematical Sociology* 17:281–302, 1993.

106. Joseph Greenberg. *The theory of social situations.* Cambridge University Press, Cambridge, 1990.

107. J. Guckenheimer and P. Holmes. *Nonlinear oscillations, dynamical systems, and bifurcations of vector fields.* Springer, New York, 1983.

108. W. Güth R. Schmittberger, and B. Schwarze. An experimental analysis of ultimatum bargaining. *Journal of Economic Behavior and Organization* 3:367–388, 1982.

109. Werner Güth and Reinhard Tietz. Ultimatum bargaining behavior: A survey and comparison of experimental results. *Journal of Economic Psychology* 11:417–449, 1990.

110. Frank Hahn. Stability. In Kenneth J. Arrow and Michael D. Intriligator, eds., *Handbook of mathematical economics,* vol. 2, pp. 745–794. North Holland, New York, 1982.

111. J. Hale. *Ordinary differential equations.* Wiley, New York, 1969.

112. Peter Hammerstein and Susan E. Riechert. Payoffs and strategies in territorial con-

tests: ESS analyses of two ecotypes of the spider *Agelenopsis aperta*. *Evolutionary Ecology* 2:115–138, 1988.

113. William Harms. Discrete replicator dynamics for the Ultimatum Game. Mimeo, University of California, Irvine, 1994.

114. J. Harsanyi. *Rational behavior and bargaining equilibrium in games and social situations*. Cambridge University Press, Cambridge, 1977.

115. John C. Harsanyi. Games with randomly distributed payoffs: A new rationale for mixed-strategy equilibrium points. *International Journal of Game Theory* 2:1–23, 1973.

116. John C. Harsanyi and Reinhard Selten. *A general theory of equilibrium selection in games*. MIT Press, Cambridge, MA, 1988.

117. Kenneth Hendricks, Robert H. Porter, and Charles A. Wilson. Auctions for oil and gas leases with an informed bidder and a random reservation price. *Econometrica* 62:1415–1452, 1994.

118. Morris W. Hirsch and Stephen Smale. *Differential equations, dynamical systems, and linear algebra*. Academic Press, New York, 1974.

119. J. Hofbauer and K. Sigmund. *The theory of evolution and dynamical systems*. Cambridge University Press, Cambridge, 1988.

120. Josef Hofbauer and Jörgen Weibull. Evolutionary selection against dominated strategies. *Journal of Economic Theory*, 1995. Forthcoming.

121. Ed Hopkins. Learning, matching and aggregation. EUI working paper ECO 95/19, European University Institute, 1995.

122. Bernardo A. Huberman and Natalie S. Glance. Evolutionary games and computer simulations. *Proceedings of the National Academy of Sciences* 90:7712–7715, 1993.

123. S. Hurkens. Learning by forgetful players. *Games and Economic Behavior* 11:304–329, 1995.

124. James S. Jordan. Bayesian learning in normal-form games. *Games and Economic Behavior* 3:60–81, 1991.

125. James S. Jordan. Three problems in learning mixed-strategy Nash equilibria. *Games and Economic Behavior* 5:368–386, 1993.

126. Ehud Kalai and Ehud Lehrer. Private-beliefs equilibrium. Center for Mathematical Studies in Economics and Management Science discussion paper 926, Northwestern University, 1991.

127. Ehud Kalai and Ehud Lehrer. Rational learning leads to Nash equilibria. *Econometrica* 61:1019–1046, 1993.

128. Michihiro Kandori, George J. Mailath, and Rafael Rob. Learning, mutation, and long-run equilibria in games. *Econometrica* 61:29–56, 1993.

129. Yuri Kaniovski and Peyton Young. Learning dynamics in games with stochastic perturbations. *Games and Economic Behavior* 11:330–363, 1995.

130. Rajeeva Karandikar, Dilip Mookherjee, Debraj Ray, and Fernando Vega-Redondo. Evolving aspirations and cooperation. Mimeo, Boston University, 1995.

131. Samuel Karlin and Howard M. Taylor. *A first course in stochastic processes.* Academic Press, New York, 1975.

132. John G. Kemeny and J. Laurie Snell. *Finite Markov chains.* Van Nostrand, Princeton, NJ, 1960.

133. Yong-Gwan Kim and Joel Sobel. An evolutionary approach to pre-play communication. *Econometrica* 63:1181–1194, 1992.

134. Motoo Kimura. *The neutral theory of molecular evolution.* Cambridge University Press, Cambridge, 1983.

135. Elon Kohlberg and Jean-François Mertens. On the strategic stability of equilibria. *Econometrica* 54:1003–1038, 1986.

136. David M. Kreps, Paul Milgrom, John Roberts, and Robert J. Wilson. Rational cooperation in the finitely repeated Prisoners' Dilemma. *Journal of Economic Theory* 27:245–252, 1982.

137. David M. Kreps and Robert J. Wilson. Reputation and imperfect information. *Journal of Economic Theory* 27:253–279, 1982.

138. David M. Kreps and Robert J. Wilson. Sequential equilibrium. *Econometrica* 50:863–894, 1982.

139. John O. Ledyard. Public goods: A survey of experimental research. In John Kagel and Alvin E. Roth, eds., *Handbook of Experimental Economics*, pp. 111–194. Princeton University Press, Princeton, NJ, 1995.

140. David K. Levine. Modeling altruism and spitefulness in experiments. Mimeo, University of California at Los Angeles, 1995.

141. D. Lewis. *Counterfactuals.* Basil Blackwell, Oxford, 1976.

142. Bruce G. Linster. Evolutionary stability in the infinitely repeated Prisoners' Dilemma played by two-state Moore machines. *Southern Economic Journal* 58:880–903, 1992.

143. Bruce G. Linster. Stochastic evolutionary dynamics in the repeated Prisoners' Dilemma. *Economic Inquiry* 32:342–357, 1994.

144. A. J. Lotka. Undamped oscillations derived form the law of mass action. *Journal of the American Chemical Society* 42:1595–1598, 1920.

145. George J. Mailath. Introduction: Symposium on evolutionary game theory. *Journal of Economic Theory* 57:259–277, 1992.

146. George J. Mailath, Larry Samuelson, and Jeroen M. Swinkels. Extensive-form reasoning in normal games. *Econometrica* 61:273–302, 1993.

147. Ramon Marimon and Ellen McGratten. On adaptive learning in strategic games. In Alan Kirman and Mark Salmon, eds., *Learning and rationality in economics*, pp. 63–101. Basil Blackwell, 1995.

148. Akihiko Matsui. Cheap-talk and cooperation in society. *Journal of Economic Theory* 54:245–258, 1991.

149. John Maynard Smith. *Evolution and the theory of games*. Cambridge University Press, Cambridge, 1982.

150. John Maynard Smith and G. R. Price. The logic of animal conflict. *Nature* 246:15–18, 1973.

151. Richard D. McKelvey and Thomas R. Palfrey. An experimental study of the Centipede Game. *Econometrica* 60:803–836, 1992.

152. A. McLennan. Justifiable beliefs in sequential equilibrium. *Econometrica* 53:889–904, 1985.

153. Paul Milgrom and John Roberts. Limit pricing and entry under incomplete information: An equilibrium analysis. *Econometrica* 50:443–460, 1982.

154. Paul Milgrom and John Roberts. Predation, reputation, and entry deterrence. *Journal of Economic Theory* 27:280–312, 1982.

155. Paul Milgrom and John Roberts. Rationalizability, learning, and equilibrium in games with strategic complementarities. *Econometrica* 58:1255–1278, 1990.

156. Paul Milgrom and John Roberts. Adaptive and sophisticated learning in normal-form games. *Games and Economic Behavior* 3:82–100, 1991.

157. Paul R. Milgrom. Good news and bad news: Representation theorems and applications. *Bell Journal of Economics* 12:380–391, 1981.

158. John H. Miller and James Andreoni. Can evolutionary dynamics explain free riding in experiments? *Economics Letters* 36:9–15, 1991.

159. Hervé Moulin. Social choice. In Robert J. Aumann and Sergiu Hart, eds., *Handbook of game theory*, vol. 2, pp. 1091–1126. North Holland, New York, 1994.

160. Roger B. Myerson. Proper equilibria. *International Journal of Game Theory* 7:73–80, 1978.

161. John H. Nachbar. "Evolutionary" selection dynamics in games: Convergence and limit properties. *International Journal of Game Theory* 19:59–89, 1990.

162. John H. Nachbar. Evolution in the finitely repeated Prisoner's Dilemma. *Journal of Economic Behavior and Organization* 19:307–326, 1992.

163. John H. Nachbar. Predition, optimization, and rational learning in games. Mimeo, Washington University, St. Louis, 1995.

164. John F. Nash. Equilibrium points in *n*-person games. *Proceedings of the National Academy of Sciences* 36:48–49, 1950.

165. John F. Nash. Two-person cooperative games. *Econometrica* 21:128–140, 1953.

166. R. Nelson and S. Winter. *An evolutionary theory of economic change*. Harvard University Press, Cambridge, MA, 1982.

167. Georg Nöldeke and Larry Samuelson. The evolutionary foundations of backward and forward induction. SFB discussion paper B-216, University of Bonn, 1992.

168. Georg Nöldeke and Larry Samuelson. An evolutionary analysis of backward and forward induction. *Games and Economic Behavior* 5:425–454, 1993.

169. Georg Nöldeke and Larry Samuelson. A dynamic model of equilibrium selection in signaling markets. SSRI working paper 9518, University of Wisconsin, 1995.

170. M. Frank Norman. *Markov processes and learning models*. Academic Press, New York, 1972.

171. Martin A. Nowak. Stochastic stategies in the Prisoner's Dilemma. *Theoretical Population Biology* 38:93–112, 1990.

172. Martin A. Nowak, Sebastian Bonhoeffer, and Robert M. May. More spatial games. *International Journal of Bifurcation and Chaos* 4:33–56, 1994.

173. Martin A. Nowak and Robert M. May. Evolutionary games and spatial chaos. *Nature* 359:826–829, 1992.

174. Martin A. Nowak and Robert M. May. The spatial dilemmas of evolution. *International Journal of Bifurcation and Chaos* 3:35–78, 1993.

175. Martin A. Nowak and Karl Sigmund. The evolution of stochastic strategies in the Prisoner's Dilemma. *Acta Applicandae Mathematica* 20:247–265, 1990.

176. Martin A. Nowak and Karl Sigmund. TIT-FOR-TAT in heterogeneous populations. *Nature* 355:250–252, 1992.

177. Martin A. Nowak and Karl Sigmund. The alternating Prisoner's Dilemma. Mimeo, University of Oxford, 1993.

178. Martin A. Nowak and Karl Sigmund. Chaos and the evolution of cooperation. *Proceedings of the National Academy of Sciences* 90:5091–5094, 1993.

179. Martin A. Nowak and Karl Sigmund. A strategy of win-stay, lose-shift that outperforms TIT-FOR-TAT in the Prisoner's Dilemma game. *Nature* 364:56–58, 1993.

180. Martin A. Nowak, Karl Sigmund, and Esam El-Sedy. Automata, repeated games, and noise. *Journal of Mathematical Biology* 33:703–722, 1993.

181. Jack Ochs and Alvin E. Roth. An experimental study of sequential bargaining. *American Economic Review* 79:355–384, 1989.

182. David Pearce. Rationalizable strategic behavior and the problem of perfection. *Econometrica* 52:1029–1050, 1984.

183. Giovanni Ponti. Cycles of learning in the Centipede Game. Mimeo, University College London, 1996.

184. Robert H. Porter. A study of cartel stability: The joint executive committee, 1880–1886. *Bell Journal of Economics* 14:301–314, 1983.

185. Philip J. Reny. Backward induction, normal-form perfection, and explicable equilibria. *Econometrica* 60:627–650, 1992.

186. Philip J. Reny. Common belief and the theory of games with perfect information. *Journal of Economic Theory* 59:257–274, 1993.

187. D. Revuz and M. Yor. *Continuous martingales and Brownian motion*. Springer, Berlin, 1991.

188. Klaus Ritzberger. The theory of normal-form games from the differentiable viewpoint. *International Journal of Game Theory* 23:207–236, 1993.

189. Klaus Ritzberger and Karl Vogelsberger. The Nash field. IAS research memorandum 263, Institute for Advanced Studies, 1990.

190. Klaus Ritzberger and Jörgen Weibull. Evolutionary selection in normal-form games. *Econometrica* 63:1371–1400, 1995.

191. Mark J. Roberts and Larry Samuelson. An empirical analysis of dynamic non-price competition in an oligopolistic industry. *RAND Journal of Economics* 19:200–219, 1988.

192. Arthur J. Robson. A biological basis for expected and non-expected utility. *Journal of Economic Theory* 68:397–424, 1996.

193. Arthur J. Robson and Fernando Vega-Redondo. Efficient equilibrium selection in evolutionary games with random matching. *Journal of Economic Theory* 70:65–92, 1996.

194. Alvin E. Roth. Bargaining experiments. In John Kagel and Alvin E. Roth, eds., *Handbook of experimental economics*, pp. 253–348. Princeton University Press, Princeton, NJ, 1995.

195. Alvin E. Roth and Ido Erev. Learning in extensive-form games: Experimental data and simple dynamic models in the intermediate term. *Games and Economic Behavior* 8:164–212, 1995.

196. Alvin E. Roth and Ido Erev. On the need for low-rationality, cognitive game theory: Reinforcement learning in experimental games with unique, mixed-strategy equilibria. Mimeo, University of Pittsburgh and Technion, 1995.

197. Ariel Rubinstein. Perfect equilibrium in a bargaining model. *Econometrica* 50:97–109, 1982.

198. Larry Samuelson. Evolutionary foundations of solution concepts for finite, two-player, normal-form games. In Moshe Y. Vardi, ed., *Theoretical aspects of reasoning about knowledge*. Morgan Kaufmann, 1988.

199. Larry Samuelson. Limit evolutionarily stable strategies in two-player, normal-form games. *Games and Economic Behavior* 3:110–119, 1991.

200. Larry Samuelson. Does evolution eliminate dominated strategies. In Ken Binmore, Alan Kirman, and Piero Tani, eds., *Frontiers of game theory*, pp. 213–236. MIT Press, Cambridge, MA, 1993.

201. Larry Samuelson. Stochastic stability in games with alternative best replies. *Journal of Economic Theory* 64:35–65, 1994.

202. Larry Samuelson and Jianbo Zhang. Evolutionary stability in asymmetric games. *Journal of Economic Theory* 57:363–391, 1992.

203. David A. Sánchez. *Ordinary differential equations and stability theory: An introduction.* Freeman, San Francisco, 1968.

204. Karl H. Schlag. Evolution in partnership games: An equivalence result. SFB discussion paper B-298, University of Bonn, 1994.

205. Karl H. Schlag. When does evolution lead to efficiency in communication games? SFB discussion paper B-299, University of Bonn, 1994.

206. Karl H. Schlag. Why imitate, and if so, how? SFB discussion paper B-361, University of Bonn, 1996.

207. Reinhard Selten. Spieltheoretische Behandlung eines Oligopolmodells mit Nach-fragetragheit. *Zeitschrift für die gesamte Staatswissenschaft* 121:301–324, 667–689, 1965.

208. Reinhard Selten. Reexamination of the perfectness concept for equilibrium points in extensive-form games. *International Journal of Game Theory* 4:25–55, 1975.

209. R. Selten. The chain-store paradox. *Theory and Decision* 9:127–159, 1978.

210. Reinhard Selten. A note on evolutionarily stable strategies in asymmetric animal contests. *Journal of Theoretical Biology* 84:93–101, 1980.

211. Reinhard Selten. Evolutionary stability in extensive two-person games. *Mathematical Social Sciences* 5:269–363, 1983.

212. Reinhard Selten. Evolutionary stability in extensive two-person games: Correction and further development. *Mathematical Social Sciences* 16:223–266, 1988.

213. Reinhard Selten. Anticipatory learning in two-person games. In Reinhard Selten, ed., *Game Equilibrium Models*. Vol. 1, *Evolution and game dynamics*, pp. 98–154. Springer, New York, 1991.

214. E. Seneta. *Non-negative matrices and Markov chains*. Springer, New York, 1981.

215. R. Seymour. Continuous-time models of evolutionary games. 1. Populations of fixed size. Mathematics working paper, University College London, 1993.

216. Lloyd Shapley. Some topics in two-person games. *Advances in Game Theory: Annals of Mathematical Studies* 5:1–28, 1964.

217. Herbert Simon. A behavioral model of rational choice. *Quarterly Journal of Economics* 69:99–118, 1955.

218. Herbert Simon. *Models of man*. Wiley, New York, 1957.

219. Herbert Simon. Theories of decision making in economics and behavioral science. *American Economic Review* 49:253–283, 1959.

220. Brian Skyrms. Chaos in game dynamics. *Journal of Logic, Language, and Information* 1:111–130, 1992.

221. Brian Skyrms. Chaos and the explanatory significance of equilibrium: Strange attractors in evolutionary game theory. Mathematical Behavior Sciences working paper 93-1, University of California, Irvine, 1993.

222. V. Smith. Experimental economics: Induced value theory. *American Economic Review* 66:274–279, 1976.

223. Joel Sobel. Evolutionary stabiliy in communication games. *Economics Letters* 42:301–312, 1993.

224. E. Somanathan. Evolutionary stability of pure-strategy equilibria in finite games. Mimeo, Emory University, 1995.

225. A. M. Spence. Job market signaling. *Quarterly Journal of Economics* 90:225–243, 1973.

226. A. M. Spence. *Market signaling*. Harvard University Press, Cambridge, MA, 1974.

227. Dale O. Stahl. The evolution of $Smart_n$ players. *Games and Economic Behavior* 5:604–617, 1993.

228. Dale O. Stahl. Rule learning in a guessing game. Center for Applied Research in Economics working paper 9415, University of Texas, Austin, 1994.

229. Paul G. Straub. Risk dominance and coordination failures in static games. Dispute Resolution Research Center working paper 106, Northwestern University, 1993.

230. Jeroen M. Swinkels. Evolutionary stability with equilibrium entrants. *Journal of Economic Theory* 57:306–332, 1992.

231. Jeroen M. Swinkels. Adjustment dynamics and rational play in games. *Games and Economic Behavior* 5:455–484, 1993.

232. Peter D. Taylor and Leo B. Jonker. Evolutionarily stable strategies and game dynamics. *Mathematical Biosciences* 40:145–156, 1978.

233. Richard H. Thaler. Anomalies: The Ultimatum Game. *Journal of Economic Perspectives* 2:195–206, 1988.

234. B. Thomas. Evolutionarily stable sets in mixed-strategist models. *Theoretical Population Biology* 28:332–341, 1985.

235. B. Thomas. On evolutionarily stable sets. *Journal of Mathematical Biology* 22:105–115, 1985.

236. Mark Twain. *A Connecticut Yankee in King Arthur's court*. In *Mark Twain: Greenwich Unabridged Library Classics*, pp. 489–690. Chatham River Press, New York, 1982.

237. Eric van Damme. A relation between perfect equilibria in extensive-form games and proper equilibria in normal-form games. *International Journal of Game Theory* 13:1–13, 1984.

238. Eric van Damme. Stable equilibria and forward induction. SFB discussion paper A-128, University of Bonn, 1987.

239. Eric van Damme. Stable equilibria and forward induction. *Journal of Economic Theory* 48:476–596, 1989.

240. Eric van Damme. *Stability and perfection of Nash equilibria*. Springer, Berlin, 1991.

241. Eric van Damme. Refinements of Nash equilibrium. In Jean-Jacques Laffont, ed., *Advances in Economic Theory: Sixth World Congress*. Cambridge University Press, Cambridge, 1992.

242. Fernando Vega-Redondo. Evolution in games: Theory and economic applications. Mimeo, University of Alicante, 1995.

243. Xavier Vives. Nash equilibrium with strategic complementarities. *Journal of Mathematical Economics* 19:305–321, 1990.

244. V. Volterra. *Leçons sur la théorie mathématique de la lutte pour la vie*. Gauthier-Villars, Paris, 1931.

245. John von Neumann and Oskar Morgenstern. *Theory of games and economic behavior*. Princeton University Press, Princeton, NJ, 1944.

246. Karl Wärneryd. Cheap talk, coordination, and evolutionary stability. *Games and Economic Behavior* 5:532–546, 1990.

247. Joel Watson. A "reputation" refinement without equilibrium. *Econometrica* 61:199–206, 1993.

248. Jürgen Weibull. *Evolutionary game theory*. MIT Press, Cambridge, MA, 1995.

249. S. Winter. Satisficing, selection, and the innovating remnant. *Quarterly Journal of Economics* 85:237–260, 1971.

250. Peyton Young. The evolution of conventions. *Econometrica* 61:57–84, 1993.

251. Peyton Young. An evolutionary model of bargaining. *Journal of Economic Theory* 59:145–168, 1993.

252. Peyton Young and Dean Foster. Cooperation in the short and in the long run. *Games and Economic Behavior* 3:145–156, 1991.

253. E. C. Zeeman. Population dynamics from game theory. In Z. Nitecki and C. Robinson, eds., *Global theory of dynamical systems*, pp. 471–497. Springer, Berlin, 1992. Lecture Notes in Mathematics No. 819.

254. E. Zermelo. Über eine Anwendung der Mengenlehre auf die Theorie des Schachspiels. *Proceedings of the Fifth International Congress of Mathematicians* 2:501–504, 1912.

Index